Reforming the Forest Service

Reforming the Forest Service

Randal O'Toole

ISLAND PRESS
Washington, D.C. • Covelo, California

About Island Press

Island Press, a nonprofit organization, publishes, markets, and distributes the most advanced thinking on the conservation of our natural resources—books about soil, land, water, forests, wildlife, and hazardous and toxic wastes. These books are practical tools used by public officials, business and industry leaders, natural resource managers, and concerned citizens working to solve both local and global resource problems.

Founded in 1978, Island Press reorganized in 1984 to meet the increasing demand for substantive books on all resource-related issues. Island Press publishes and distributes under its own imprint and offers these services to other nonprofit organizations.

Funding to support Island Press is provided by The Mary Reynolds Babcock Foundation, The Educational Foundation of America, The Charles Engelhard Foundation, The Ford Foundation, The George Gund Foundation, The William and Flora Hewlett Foundation, The Joyce Foundation, The J.M. Kaplan Fund, The John D. and Catherine T. MacArthur Foundation, The Andrew W. Mellon Foundation, The Joyce Mertz-Gilmore Foundation, Northwest Area Foundation, The Jessie Smith Noyes Foundation, the J.N. Pew, Jr. Charitable Trust, The Rockefeller Brothers Fund, The Florence and John Schumann Foundation, and The Tides Foundation.

For additional information about Island Press publishing services and a catalog of current and forthcoming titles, contact Island Press, P.O. Box 7, Covelo, California 95428.

Library of Congress Cataloging-in-Publication Data

O'Toole, Randal Lee
Reforming the Forest Service.

Notes, graphs.
Includes index.
1. West (U.S.)—Public lands.
ISBN 0-933280-49-1
ISBN 0-933280-45-9 (pbk.)

Manufactured in the United States of America
95 94 93 92 91 90 89 5 4 3

Third Printing

Contents

List of Figures

List of Tables

Preface

In the past five years, my work as a forest economist for CHEC — a nonprofit forestry consulting firm that works almost exclusively for conservation groups — has been a source of increasing dismay and frustration. I've visited national forests in every part of the country and have seen costly environmental destruction on a grand scale. Money-losing timber sales are costing taxpayers at least $250 to $500 million dollars per year. Many of these sales are reducing scarce recreation opportunities, driving wildlife species toward extinction, and polluting waters and fish habitat. Yet few in the Forest Service seem to be concerned about these problems except as they affect public relations.

Only recently did I realize that this lack of concern was due not to ignorance or maliciousness but rather to a lack of any incentive to be concerned. Most of the Forest Service's budget comes straight from Congress. Unlike a private company, whose job it is to keep stockholders happy by producing profits, the Forest Service must keep Congress happy by creating jobs and income for local constituents. Since neither the Forest Service managers, the local constituents, nor the members of Congress have to pay for those jobs — which often cost far more than the workers are paid — none of them has an incentive to compare expenses with income.

In addition, a large chunk of the Forest Service's budget — approaching $300 million per year — is retained by agency managers out of timber sale receipts. This income gives managers a powerful incentive to sell timber, since even timber that loses money can contribute to the managers' own budgets. This positive feedback from timber sales also gives managers little incentive to compare the total expenses with the total income of those sales.

Despite the lack of interest within the Forest Service, however, controversies over environmental problems, combined with the fact that deficit spending often accompanies environmental destruction, have led many outside the Forest Service to compare timber expenses and income. Many studies have shown that Forest Service timber sales lose money, but few have ever asked why that is so. The answer to this question has important implications for any attempts to reform the Forest Service — or any other natural resources agency.

Reforming the Forest Service is an outgrowth of my reviews of more than forty national forest plans and thousands of timber sales. My economic research has convinced me that Americans can have all the wilderness, timber, wildlife, fish, and other forest resources they want. Apparent shortages of any of these resources are due solely to the Forest Service's failure to sell them at market prices.

However, Americans are *not* likely to get what they want from the national forests under the current system of forest planning and management. The interest group politics that dominates public forest management guarantees that the environmental quality of the national forests will continue to deteriorate, at taxpayers' expense. Major legislative reforms in national forest management — indeed, thorough reforms of the Forest Service as an institution — are needed to correct this. The reforms proposed in this book will, if implemented, prevent resource shortages; eliminate below-cost sales of timber, grazing, and other resources; and put an end to most of the divisive and polarized environmental conflicts that have plagued the Forest Service for the past thirty years.

While the proposals are not entirely new, they require a radical rethinking of traditional environmental concepts. As one who has worked primarily for environmentalists for nearly fifteen years, I hope that environmentalists will approach this book with an open mind and a willingness to apply the concepts described here to their own experiences.

Although the proposals in this book are fairly detailed, they are far from complete. Over the next two years I plan to do additional research on Forest Service behavior, projecting the effects of the reforms on individual forests and finding examples of land management similar to the proposals. The results of this work will be published in CHEC's magazine, *Forest Watch*, and possibly in a future edition of this book.

It is appropriate here to thank Lisa Schwartz and Nancy Boulton for their thorough reviews of the manuscript, David O'Toole for designing and laying out the book, Ken Fletcher for helping prepare the index, and Vickie Crowley for much advice and counsel. I'd also like to thank Richard Alston, John Baden, Bill Ferrell, Jerry Franklin, Dave Iverson, Al Sample, and Andy Stahl for commenting on the concepts and proposals in early drafts of the book. Naturally, I alone am responsible for the data, conclusions, and proposals.

Randal O'Toole
Eugene, Oregon
September 1987

Introduction
A Tale of Three Forests

The Tale of the Santa Fe National Forest

Early in 1986 I join Ted Davis in a hike on Virgin Mesa, located near Jemez Canyon in the Santa Fe National Forest. Davis, a soft-spoken physician from Albuquerque, leads me to an Indian pueblo ruin called Kwastiyukwa. Although the site is no farther from Interstate 25 than popular Bandelier National Monument, Kwastiyukwa is unknown to the tens of thousands of tourists who visit New Mexico to see Bandelier and other Indian ruins each year.

Yet this former home of hundreds of Indians is an impressive sight. The remains of many rooms and several *kivas* — the circular rooms used in pueblo religious ceremonies — are clearly evident. From the air, Kwastiyukwa's kivas and pueblos form an outline of a large footprint, and the ruin is sometimes known as "bigfoot." On the ground, we can see fingerprints in the exposed adobe, the marks of an ancient artisan. According to Davis, Kwastiyukwa is only one of several ruins in the vicinity. Another, known as Tovakwa, once housed 2,000 people and is so huge — nearly 1,000 feet by 1,200 feet — that the famous pueblo ruin at Chaco Canyon could fit entirely within its plaza.

Our hike takes us to the edge of the mesa, where we see the remains of an ancient Indian trail that led from the valley to the pueblo. From this vantage point and all along the hike, we also look back at the trail markings of another people: daubs of yellow paint. Splashes of this paint are seen on the stunted, bonsai-like trees leaning over the cliff, and more yellow paint adorns the boulders that we climb over and step around.

What does this yellow paint mean? "The Forest Service is planning a timber sale on the mesa," says Davis. "The paint indicates all the remains of Indian habitation that the loggers are supposed to avoid." With so much yellow paint around, it seems incredible that even a single tree could be removed without disturbing one of the cultural sites.

The Forest Service wasn't always so careful with its timber sales. Davis became an activist in forest management and timber sale planning primarily because of his concern that the managers of the Santa Fe National Forest were destroying priceless cultural resources to grow and sell trees. Davis leads Save the Jemez, a local coalition of archaeologists, native Americans, and environmentalists interested in protecting areas such as Virgin Mesa from the chain saw.

In 1983, a freedom-of-information request filed by Save the Jemez forced the Forest Service to reveal numerous violations of the Antiquities Act caused by timber management activities. "The Forest Service was using a plow to prepare sites for reforestation," explains Davis. "The plow had disks that cut 18 inches into the soil and destroyed any signs of Indian habitation." According to the documents obtained under the Freedom of Information Act, Forest Service archaeologists estimated that repairing the damage caused by such site preparation activities in several locations would cost hundreds of thousands of dollars.[1]

At about the same time Davis received this information, the Santa Fe National Forest was distributing its final forest plan to the public. As required by the National Forest Management Act of 1976 (NFMA), the Santa Fe and 117 other national forests were publishing forest plans that allocated land to various uses, such as timber and recreation; outlined management standards to ensure that logging, grazing, and other activities would not harm other resources; and fixed the "allowable sale quantity," or the maximum amount of timber that could be sold each year. The Santa Fe Forest had been selling about 40 million board feet of timber each year (a board foot is 1 inch by 12 inches by 12 inches of lumber). But the plan proposed to increase this to 46 million board feet for the next ten years, possibly reaching as high as 74 million board feet within the next 30 years.[2]

"Santa Fe Forest archaeologists estimate that there are 27,000 cultural resource sites on Santa Fe Forest land," Davis tells me. "Most are Indian ruins. Logging almost inevitably damages or destroys these sites." With so many sites, the destruction of a few may seem unimportant. After all, timber is a valuable resource, and the Forest Service can't be expected to halt all timber production just for a few Indian ruins. This may have been the Forest Service managers' reasoning as they plowed up the hillsides and planned new timber sales in the heart of the richest concentration of Indian ruins in the nation. Yet the basic premise behind this reasoning is flawed: Timber grown in the Santa Fe National Forest is not a valuable resource.

The Forest Service has been selling timber from the Santa Fe Forest for years, and a number of companies actively bid on the timber in oral auctions. Despite this competition, the Santa Fe has continually spent more money on timber management than it collected in sale receipts — even in the late 1970s, when timber prices reached an all-time high. Between 1974 and 1978, for example, timber management and road construction costs exceeded timber receipts by more than 10 percent.[3]

"Many of the areas in the Santa Fe are still sacred to our people," says Jose Lucero, a Santa Clara Indian who is the former director of the New Mexico Soil and Water Conservation Division. Lucero, who has close ties to the traditional religious community of his tribe, is an active member of Save the Jemez. "Why does the Forest Service destroy these areas if it loses money?"

"Recreation is a growing industry in New Mexico," Davis points out. "Many of these ruins could be a major attraction and contribute far more to the economy than timber sales. Why doesn't the Forest Service understand this?" he asks.

The Tale of the Beaverhead National Forest

Similar questions are being asked by Jim Welch, a cattle rancher in Wisdom, Montana. Born on a ranch in New Mexico, Welch left home early and made a fortune in the southern California construction industry. Then he decided to return to ranching, and he bought the Arrow Ranch in the Big Hole, an isolated valley in southwestern Montana. Behind Welch's ranch, the Anaconda-Pintlar Wilderness teems with wildlife. Elk, moose, deer, bighorn sheep, and perhaps even a grizzly bear can be seen during a walk through the area. But between the Arrow Ranch and the wilderness is a strip of undesignated forestland — land that the Forest Service manages for "multiple use." In 1983, Welch learned that the Beaverhead National Forest planned to build a road across this strip in order to sell the timber.

According to Forest Service estimates, the timber was worth far less than it would cost to build the road. But the road was necessary, said officials, to provide recreationists with access to the wilderness. The fact that a trail could provide access at a fraction of the cost was considered irrelevant: With circular reasoning, the Forest Service pointed out that the road would eventually be necessary to sell the trees, so it might as well be built now.

"The Forest Service also claims that cutting timber will benefit the wildlife," says Welch. "One day, a group of Forest Service officials and I hiked through the area. As we entered one corner of a large, natural meadow, we saw a black bear and cub across the meadow. In another corner we could see a cow moose and calf. And on the other side were several elk. How, I asked the Forest Service officials, could a timber sale improve on that?"

The Montana Department of Fish, Wildlife, and Parks agrees that conditions for game and other wildlife are nearly optimal in the Big Hole and that both roads and timber harvests are likely to reduce game habitat. "Increased road construction must be balanced against proposed [i.e., projected] decreases in elk hunting quality and potentially greater restrictions on hunting opportunities," the department told the Forest Service.[4] Departmental biologists noted that proposed timber sales in the Big Hole would cause "a major reduction in elk security habitat."[5]

On another day in late spring 1985, Welch and I ride horseback over Beaverhead National Forest land. We dismount in a lush, grassy meadow and look at the clear stream bubbling nearby. Although it is a warm day, the water is icy. But Welch rolls up his shirt sleeve and reaches down into the water. "This is called Clam Creek," he says as he rises, holding a few freshwater clams. "These clams are extremely sensitive to sedimentation. A network of roads up here would wipe them out." Montana law forbids degradation of water quality in streams such as Clam Creek. Although the Forest Service claims it will obey this law, it refuses to consider whether road construction would harm the clams. "After I kept badgering them about the clams," says Welch, "one Forest Service official finally told me that 'no one cares about your gud-damned clams!'"

Perhaps no one does care about the clams. But many people care about the fish that live and grow in Clam Creek, Thompson Creek, and other headwater streams of the Big Hole River. The Big Hole is a "blue ribbon" fishery: According to the

Montana State Department of Fish, Wildlife and Parks, "The upper Big Hole River and its tributaries support the last native stream dwelling population of grayling [trout] in the United States south of Alaska."[6]

Fish are a valuable resource in the Beaverhead Forest, introducing nearly $250,000 dollars into the Montana economy each year.[7] Yet the Big Hole, which is the most pristine drainage in the forest, is targeted for heavy timber cutting in the next ten years. The valley contains only 30 percent of the forest, but plans call for over 60 percent of the timber cutting and over 70 percent of new road construction to take place there.[8] The Forest Service also wants to build roads into thousands of acres of roadless areas to provide access to more timber.

Although timber cutting threatens wildlife, fish, and primitive recreation — all valuable resources that cost relatively little to provide — the timber itself is essentially worthless. Composed mostly of lodgepole pine and Rocky Mountain Douglas-fir — a puny cousin of the Douglas-fir found west of the Cascades in the Pacific Northwest — Forest Service timber often sells for as little as $5 to $10 per thousand board feet and rarely brings over $100 per thousand board feet. Thus, the economic return from Beaverhead timber sales is minimal.

Between 1980 and 1986, the Forest Service collected $4.3 million for timber sales from purchasers of Beaverhead Forest timber. The term *collected* is used with caution, however. Under a 1964 law, the Forest Service can require purchasers to build the roads needed to reach the timber. The cost of those roads can then be credited against the value of the timber. For this reason, such roads are often called "purchaser credit" roads. Of the $4.3 million paid by purchasers for Beaverhead timber, $1.7 million was paid in the form of roads built by timber purchasers.

Under the Knutson-Vandenberg Act of 1930, the Forest Service can keep from timber sale receipts the amount of money it estimates is needed to reforest timber sale areas. The National Forest Management Act expanded the use of these funds to include thinnings, wildlife habitat improvement, and other activities. The so-called "K-V" monies are deposited in a trust fund until they are actually spent on forest management. On the Beaverhead Forest, $2.0 million of the $4.3 million was retained by the Forest Service for K-V activities.

An additional $1.1 million, or 25 percent of the total, was turned over to counties under the provisions of the Act of May 23rd, 1908, whereby the Forest Service gives one-fourth of all its receipts — including K-V deposits and purchaser road credits — to the counties for roads and schools. Although this money was originally intended to substitute for the taxes that would be paid if the land were privately owned, in fact many counties receive far more from the Forest Service than they would collect in taxes.

These three amounts — $1.7 million for roads, $2.0 million for the K-V Fund, and $1.1 million for counties — add up to over $400,000 *more* than the total receipts collected from 1980 to 1986, an average of $60,000 per year. Thus, the U.S. Treasury lost money on Beaverhead timber sales before it even spent a dime on timber management.[9]

Yet it did spend a dime, and quite a bit more, on Beaverhead timber activities. Between 1980 and 1985, the Beaverhead Forest typically spent $360,000 per year just to arrange and administer timber sales. It also spent $800,000 per year on road construction and engineering and design of purchaser credit roads. In addition, the forest spent over $200,000 per year on reforestation and precommercial thinning above and beyond the amount it spent from the K-V Fund. More money was spent on related costs, such as locating land lines so timber cutters would not trespass onto private lands, maintaining nurseries for seedlings, and maintaining timber roads.[10]

Thus, at a cost of over $2 million per year, the U.S. Treasury earned the privilege of collecting a negative $60,000 per year in timber sale receipts. The 1980s were not banner years for the timber industry, but the Beaverhead Forest failed to earn a positive return at any time during the late-1970s, which were profitable years for the industry.

"The Big Hole River is a sporting paradise," Welch says. "It's world famous as a fishery, and it produces abundant wildlife. Why does the Forest Service want to destroy it by selling timber if timber loses money?"

The Tale of the Nantahala-Pisgah National Forests

In Asheville, North Carolina, two thousand miles from Montana and New Mexico, people ask the same question. Asheville was the adopted home of George Vanderbilt, scion of the railroad builders. Here he constructed Biltmore, the famous castlelike home from which he administered an estate that spanned tens of thousands of acres. Here, too, he hired Gifford Pinchot to develop the first "scientifically managed" forest in the nation.

The son of a wealthy Pennsylvania family, Pinchot went to Europe to study forestry because at that time there were no forestry schools in the United States. On his return from France, one of Pinchot's first jobs was to manage the Biltmore Forest. Pinchot rejected clearcutting, which was practiced — carelessly, he believed — by many timber companies. Instead, he selectively managed the forest, choosing trees to harvest with care to ensure that the forest would remain healthy and productive.

Pinchot left his position at Biltmore to take over the Bureau of Forestry, in the U.S. Department of Agriculture. In 1903, Pinchot succeeded in getting the nation's forest reserves placed under his jurisdiction. To distinguish his agency from the many other bureaucracies, he renamed it the Forest Service. A favorite of Teddy Roosevelt, Pinchot and his Forest Service grew in power and in the number of acres they managed.

George Vanderbilt died in 1914, and his heirs agreed to turn Biltmore Forest over to Pinchot's Forest Service. In a gesture that must have been painful to Pinchot, however, they retained the timber rights and sold them to a local timber company. In a few years, Biltmore Forest was clearcut from end to end.

Today, most of the former Biltmore Forest has grown back, and it is now part of the Pisgah National Forest which is particularly revered by the Forest Service as the "Cradle of Forestry." The Pisgah Forest, together with the Nantahala, the Uwharrie, and the Croatan forests, all located in North Carolina, are managed by one forest supervisor's office in Asheville.

The Uwharrie and Croatan forests are located on the eastern seaboard of the state. Highly productive timberlands, they are relatively flat and uninteresting to most recreationists. The Nantahala and Pisgah forests, however, are located in the southern Appalachian mountains. In close proximity to Great Smoky Mountains National Park, the forests are traversed by millions of sightseers, who particularly enjoy the Blue Ridge Parkway during the fall "color" season. Hunting, fishing, and hiking are also enjoyed there.

Jim Dockery and Bob Smythe, of the North Carolina Chapter of the Sierra Club, are interested in the recreation potential of the Nantahala and Pisgah forests. "We don't have large wilderness areas here like you do out West," Dockery tells me in a pleasant drawl. "But we do have some smaller roadless areas that we would like to protect for backcountry recreation."

The Forest Service had other ideas. With a road network planned for almost 90 percent of the forests, only the Congressionally designated wilderness and a few areas reserved for research, the Appalachian Trail, and similar uses would remain unmarked by roads.[11] Other roadless areas the Sierra Club wanted to protect were nearly all scheduled for roads and logging. Yet the timber values were too low to justify logging every tree — or even most trees. A computer calculation made by the Forest Service to determine the highest value of the forests advised that nothing but Christmas trees should be cut for at least fifty years, when timber prices were optimistically expected to rise high enough to cover logging and road costs.[12] Forest Service managers ignored this calculation and, in fact, would probably not have made the calculation if it had not been for the curiosity of the computer programmer.

The Forest Service claims that the Nantahala and Pisgah forests are poorly managed, in part the result of careless clearcutting practices early in this century. The second-growth forests that will be grown after harvests today, the Forest Service promises, will be much more productive and will return far more revenue. Yet the investments the Forest Service plans to make in reforestation and timber management are likely to return, at best, 3.5 percent per year, and most investments are expected to produce 1 percent or less.[13] The government could get a better return and could grow more timber by investing in private forestlands, which are more productive than most national forests.

"National forest timber accounts for only about 5 percent of all the timber cut in this area," Smythe says. "No one would ever notice if the Forest Service stopped selling it. Why does it want to sell so much timber when it loses so much money?"

The Tale of 156 National Forests

From Alaska to Georgia, from California to Vermont, people who are concerned about the national forests have been asking this question. In the face of mounting evidence that almost all of the 156 national forests lose money on at least some timber sales, the Forest Service doggedly denies that there is a problem and insists that more, not less, timber should be sold from each forest. In the face of growing criticism from hunters, recreationists, ranchers, and other forest users, the Forest Service maintains its assumption that timber sales are needed to benefit all other forest uses.

Ted Davis, Jim Welch, and Jim Dockery have all found that Forest Service managers are responsive to concerted citizen action. Through lawsuits and appeals, Davis has made the Santa Fe Forest more sensitive to cultural resources. Using the media and promoting write-in campaigns, Welch has convinced the Forest Service to withdraw from timber management national forestlands adjoining the Anaconda-Pintlar Wilderness. Through public involvement in the forest planning process, the North Carolina Chapter of the Sierra Club has persuaded the North Carolina forest supervisor, George Olson, to protect more roadless lands from development.

Yet money-losing timber sales continue in all of these and many other national forests. Such sales will no doubt damage cultural resources, destroy fish and wildlife habitat, and reduce the opportunities for high-quality primitive or semi-primitive recreation. The successes achieved to date are small compared to the changes needed to put the Forest Service on a sound economic and environmental footing.

What makes the Forest Service so intractable and apparently irrational? The answer to this question is key to any reform of the Forest Service because reforms that treat the symptoms and not the causes of Forest Service problems are doomed to fail. This book will show that inefficient management and environmental controversies are not problems in themselves but are merely symptoms of major institutional defects within the Forest Service. Because of these defects, actions that appear to an outsider to be irrational, inefficient, and environmentally destructive are, in fact, perfectly reasonable from the viewpoint of the Forest Service.

Of course, the Forest Service is not the only federal agency in need of reform. The Bureau of Land Management, the Bureau of Reclamation, the Corps of Engineers, and many other agencies repeatedly support policies that are destructive to the environment and economically wasteful. If the reforms proposed in this book are successful in improving national forest management, similar reforms could be designed and applied to many other agencies. The result could protect many environmental values as well as save the U.S. government billions of dollars each year.

Reform of an agency with an eighty-year history, especially one as large and well respected as the Forest Service, is an ambitious goal. But twelve years of studying the agency in detail have convinced me that without such reform, the Forest Service will continue to degrade the quality of America's environment and make major contributions to the national deficit.

The time is ripe for reform. Environmentalists are institutionalized as a powerful lobby in Washington, DC, yet they maintain active grassroots support. At the same time, fiscal conservatives are growing in strength as people recognize that government spending may be more harmful than helpful. The reforms proposed by this book should also be supported by local governments since the proposals will actually increase national forest payments made to most counties in lieu of taxes. Private landowners will benefit as well because a halt to Forest Service subsidies will make their lands more valuable. Together, these potential beneficiaries can work towards the mutual goal of Forest Service reform.

By itself, the Forest Service has been unable to resolve a number of major controversies including below-cost sales, sustained yield, wilderness, clearcutting, herbicides, and grazing. The most striking of these is below-cost sales because timber is sold in competitive markets, which theoretically should return a fair value for the products sold. After a brief look at the history and organization of the Forest Service, part I of this book examines below-cost sales in detail, showing that few, if any, of these sales are rational from a social, economic, or environmental viewpoint.

Part II shows that the Forest Service persists in selling timber at a loss, as well as maintaining other controversial and seemingly irrational policies, because that behavior is in its own interest. Part III shows that the national forests are doomed to suffer costly and environmentally destructive management as long as the current system of interest group politics exists. Reforms are designed to make the Forest Service's self-interests coincide, as much as possible, with public demands for national forest resources. The resulting legislation should greatly improve the environmental quality of the national forests and, as a bonus, save federal taxpayers more than $2 billion per year.

Notes

1. "Archeological Crew Training on Holiday Mesa," memo on file at the Santa Fe National Supervisor's Office, Santa Fe, NM, 1981, 6 pp.
2. USDA Forest Service, *Santa Fe Forest Plan Final Environmental Impact Statement* (Santa Fe, NM: Forest Service, 1983), p. 35.
3. Thomas J. Barlow et al., *Giving Away the National Forests* (Washington, DC: Natural Resources Defense Council, 1980), appendix 1.
4. USDA Forest Service, *Beaverhead Forest Plan Final EIS* (Dillon, MT: Forest Service, 1986), p. IV-79.
5. "Wildlife," draft comments on Beaverhead supplemental DEIS prepared by Montana Department of Fish, Wildlife, and Parks, p. 6.
6. "Fisheries," draft comments on Beaverhead supplemental DEIS prepared by Montana Department of Fish, Wildlife, and Parks, p. 3.
7. USDA Forest Service, *Beaverhead Final EIS*, p. B-72.
8. CHEC, "Review of the Beaverhead National Forest Draft Plan and EIS" (Eugene, OR: CHEC, 1985), p. 13.
9. USDA Forest Service, *Region 1 Annual Collections Statements [for] Fiscal Years 1980-86* (Missoula, MT: Forest Service, 1981-1987).
10. USDA Forest Service, *Region 1 Annual Budget Summaries [for] Fiscal Years 1980-86* (Missoula, MT: Forest Service, 1981-1987).
11. USDA Forest Service, *Nantahala-Pisgah Forest Plan Draft Environmental Impact Statement* (Asheville, NC: Forest Service, 1984), p. II-76.
12. CHEC, "Review of the Nantahala-Pisgah National Forests Draft Plan and EIS" (Eugene, OR: CHEC, 1984), p. 6.
13. Ibid., p. 5.

Part I

Analyzing the Symptoms

Chapter One
An Economic Contradiction

The Forest Service is the largest agency in the United States Department of Agriculture, employing over 30,000 people and managing over 191 million acres of land.[1] The national forests, which it manages, sell between 10 and 11 billion board feet of wood each year, provide grass for millions of cattle and sheep, host millions of recreationists, and are the source of water for numerous cities. Although the Forest Service claims that the national forests produce over $3 billion in benefits each year, actual receipts fall short of expenses by more than $1 billion per year.

In 1985, the agency received an average of $85 per thousand board feet for timber cut from the forests (table 1.1).[2] However, the Forest Service spent close to $89 per thousand board feet preparing and administering sales, building timber access roads, and reforesting and managing new stands of timber (table 1.2).

The forests also sell about 10 million animal unit months (AUMs) of forage for domestic livestock each year. By law until 1986 and more recently by executive

Table 1.1 1985 and 1986 Timber Receipts in Millions of Dollars

Item	*1985*	*1986*
Timber Account	$498.5	$726.0
K-V Deposits	186.1	151.5
Purchaser Credits	107.9	117.0
Timber Salvage Sales	15.2	20.7
Brush Disposal	53.7	52.9
Cooperative Work	39.9	46.4
O&C Lands	16.1	19.1
Total Timber Receipts	917.4	1,133.6
Payment to Counties	210.0	263.4
Retained by Forest Service	402.8	388.5
Net to Treasury	$304.6	$481.8

About 10.9 billion board feet were cut from the national forests in 1985 and about 11.8 in 1986. Although total receipts exceeded $900 million in both years, the U.S. Treasury collected less than half of that amount.

Source: USDA Forest Service, *1987 Budget Explanatory Notes* (Washington, DC: Forest Service, 1986), p. 11; USDA Forest Service, *1988 Budget Explanatory Notes* (Washington, DC: Forest Service, 1987), p. 13.

Table 1.2 1985 and 1986 Timber Expenses in Millions of Dollars

	1985	*1986*
Appropriations		
Timber Sales	$184.5	$174.0
Resource Support	53.8	29.1
Reforestation & TSI	145.9	95.1
Road Construction	234.8	157.8
Tongass Timber Fund	47.0	45.8
Total Appropriations	$666.0	$501.8
Funding out of Receipts		
K-V Reforestation & TSI	$90.0	$85.8
Brush Disposal	41.8	46.0
Purchaser Credits	192.3	91.5
Timber Salvage Sales	16.1	23.0
Total Funds out of Receipts	$340.2	$246.3

Timber-related costs exceeded receipts in 1985 but were exceeded by receipts in 1986. But over $500 million was spent from appropriated funds in both years, well over the amount actually returned to the Treasury (see table 1.1).

Source: USDA Forest Service, *1987 Budget Explanatory Notes* (Washington, DC: Forest Service, 1986), p. 150–157; USDA Forest Service, *1988 Budget Explanatory Notes* (Washington, DC: Forest Service, 1987), p. 155.

order, the agency charges only $1.35 per AUM for most of this grazing, but Forest Service economists estimate that its true value is closer to $5 per AUM. The Forest Service spent $2.79 for each AUM in 1985 but actually received an average of only $0.77.[3]

The national forests provide twice as much outdoor recreation as the national parks and many times as much outdoor recreation as Bureau of Land Management lands even though the latter cover a greater area. In 1985, millions of Americans spent 225 million recreation visitor days — a Forest Service unit of measure that counts a twelve-hour visit by one person as a day — in the national forests. Forest Service economists estimate that these visitor days are worth $4 to $30 or more. However, actual national forest receipts in 1985 averaged only $0.14 per visitor day, while the Forest Service spent $0.45 per visitor day on recreation — most of which was used for administration.[4]

An additional 32 million visitor days were spent hunting and fishing on Forest Service lands. Half the big game and coldwater fish populations in the U.S. are thought to occur in the national forests, which also provide unique habitat for many endangered species. Although economists estimate that hunting and fishing are the most valuable forms of outdoor recreation, national forest receipts for this activity are only incidental. Yet the agency spent over $36 million on fish and wildlife in 1985, more than $1 per visitor day.[5]

The National Forest System is also the source of drinking water for millions of Americans. National forest watersheds are the principal source of water for Seattle, Portland, Denver, Los Angeles, and many other cities. Billions of acre-feet of water flow from the national forests each year, much of it used for municipal purposes, and

Table 1.3 National Forest Resource Values and Volumes in 1985

Resource	*Volume*	*Value*	*Receipts*	*Cost*
Timber	10.9	$919.4	$917.4	$965.1
Grazing	10.1	56.0	9.0	32.1
Recreation	228.0	1,393.0	30.8	121.0
Fish & Wildlife	32.0	400.0	0.0	36.7
Water	n.a.	81.0	0.0	31.8
Minerals	n.a.	585.0	160.5	26.6
Total		$3,434.4	$1,117.7	$1,213.3

Timber volumes are in billions of board feet, grazing in millions of animal unit months, and recreation and fish and wildlife in millions of recreation visitor days. All dollar values are in millions.

Source: Volumes, receipts, and timber costs in USDA Forest Service, *1987 Budget Explanatory Notes* (Washington, DC: Forest Service, 1986), nontimber values in USDA Forest Service, *A Recommended Renewable Resources Program: 1985-2030* (Washington, DC: Forest Service, 1986), p. 12. The difference between timber value and timber receipts is due to firewood removals as indicated in the *RPA Program*, p. 12.

much more used for irrigation. Although the Forest Service claims that irrigation water is worth $12 to $59 per acre-foot, and municipal and industrial water is worth much more, the national forests receive no income for the water they produce. Yet the agency spent over $30 million on watershed management in 1985.[6]

In 1986, *Fortune* magazine reported that the top five U.S. industrial corporations in terms of their 1985 assets were Exxon ($69.2 billion), General Motors ($63.8 billion), IBM (52.6 billion), Mobil ($41.8 billion), and AT&T (40.5 billion).[7] If national forest assets are between $42 and $50 billion, the Forest Service would outrank Mobil to place fourth on this list.

Total national forest receipts in 1985 were $1.12 billion. This would place the Forest Service between Inspiration Resources, Fortune industrial number 289, and CertainTeed, Fortune 290.

The national forests thus represent a major economic contradiction. They pro-

Table 1.4 Disposition of 1985 and 1986 Forest Receipts

	1985	*1986*
Total Returns	$1,132.7	$1,315.9
Retained by Forest Service	306.4	289.4
Payments to Counties	235.7	497.6
Transfer to USDI	98.5	96.8
Noncash Receipts	107.9	117.0
Returns to Treasury	$384.2	$315.1

All values are in millions. Although the Forest Service reported receipts exceeding $1.1 billion in both years, the Treasury collected less than $400 million.

Source: USDA Forest Service, *1987 Budget Explanatory Notes* (Washington, DC: Forest Service, 1986), p. 11; USDA Forest Service, *1988 Budget Explanatory Notes* (Washington, DC: Forest Service, 1987), p. 13.

Table 1.5 Forest Service Appropriations in Millions of Dollars

	1985	*1986*
Research	$113.8	$120.1
State & Private Forestry	58.3	55.3
National Forests	1,111.5	1,168.9
Construction	263.9	214.7
Aquisitions	51.3	33.2
Trust & Work Funds	433.9	390.3
County Payments	235.7	497.6
Total Cost	$2,268.4	$2,480.1
Cost to Treasury	1,598.8	1,592.2

The Forest Service typically spends over $2.2 billion per year, of which nearly $1.6 comes from appropriations. Most trust and work funds and some construction funds are paid out of receipts.

Source: USDA Forest Service, *1987 Budget Explanatory Notes* (Washington, DC: Forest Service, 1986), p. 12; USDA Forest Service, *1988 Budget Explanatory Notes* (Washington, DC: Forest Service, 1987), p. 14.

duce over $3.4 billion worth of resources each year (table 1.3),[8] yet total returns to the U.S. Treasury after payments to counties in lieu of taxes were barely $380 million in 1985 (table 1.4).[9] To produce this $380 million return, the Forest Service receives over $1.5 billion per year in congressional appropriations for national forest management (table 1.5).[10] In terms of assets, the agency would rank in the top five in *Fortune* magazine's list of the nation's 500 largest corporations. In terms of operating revenues, however, the agency would be only number 290. In terms of net income, the Forest Service would be classified as bankrupt.

In 1976, Marion Clawson of Resources for the Future estimated that national forest assets were worth approximately $42 billion in 1974.[11] Almost half of this was Clawson's estimate of the value of timber on the forests. Although nominal timber prices increased considerably between 1974 and 1980, they have fallen just as dramatically since 1980. At today's timber prices in today's dollars, a recalculation of Clawson's value would produce a number less than $50 billion.

In 1985, the Fortune 500 company with the largest reported loss was Ling-Tempco-Vought, which lost $725 million. Ford Motor Company teetered on the edge of bankruptcy after losing over $1 billion per year for two years running in 1980-81. More than two or three years of such losses would cause any company to fail. Yet congressional appropriations to the Forest Service continually exceed returns to the Treasury by well over $1 billion. Although the agency claims receipts will exceed expenses by the year 2000 if timber prices increase as it predicts, it makes no provision for reducing expenses if prices do not increase.[12]

Given assets of $42 billion, why does the agency cost the U.S. Treasury over $1 billion per year? The Forest Service says that "treasury receipts are not a good estimate of the total value of the National Forest System to society since only those outputs that are actually sold are included."[13] Most recreation, wildlife, and water benefits, in particular, are not sold. Grazing is sold, but, as mandated by the president, prices are well under fair market value.

To test the Forest Service's claim, Clawson prepared a financial statement for the National Forest System as of 1974, including interest on assets and noncash benefits. Considering only direct expenditures and income, the Forest Service spent a few million dollars more than it collected in 1974. Clawson modified revenues by adding the value of roads built by timber purchasers and recreation consumed without charge by the public. He adjusted expenses by adding depreciation, payments to counties, and 5 percent interest on all assets. When these other income and expense items were considered, Clawson found that costs exceeded benefits by nearly $2 billion (table 1.6).[14]

The 5 percent interest on assets made the difference between positive and negative net value. With $42 billion worth of assets, interest totals to $2.1 billion. Clawson's point was that in failing to use standard capital accounting principles, the Forest Service was doing a disservice to taxpayers. If the national forests were to be sold — a possibility that Clawson did not seriously suggest at the time — and the sale receipts invested at 5 percent, taxpayers would obtain a greater return. Clawson's work suggests that the national forests are being inefficiently managed on a large scale.

Repeating this calculation for 1985 indicates that the Forest Service lost only $1.2 billion in that year. The improvement is due not to net receipts, which have actually declined, but to an increase in the assumed value of noncash benefits. Clawson estimated that recreation was worth $2 per visitor day, wilderness recreation was worth $10 per visitor day, and water produced by the national forests in

Table 1.6 National Forest Income Statement for 1974 and 1985

	1974	*1985*
Income		
In Cash	$486.0	$839.5
In Kind	220.0	201.5
Non-Cash	490.0	1,893.0
Total	$1,196.0	$2,934.0
Expenses		
In Cash	$488.0	$1,515.4
In Kind	220.0	201.5
Depreciation	200.0	200.0
Payments to States	79.0	235.7
Interest on Assets	2,100.0	2,100.0
Total	$3,087.0	$4,252.6
Net Benefits	-$1,891.0	-$1,318.6

Clawson's 1974 income statement and CHEC's 1985 update using RPA data for recreation and wildlife values shows all values in millions of dollars.

Source: USDA Forest Service, *1987 Budget Explanatory Notes* (Washington, DC: Forest Service, 1986), pp. 11-12, and Marion Clawson, *The Economics of National Forest Management* (Baltimore, MD: Johns Hopkins University Press, 1976), p. 56.

1974 was worth $50 million. The Forest Service now assumes that recreation is worth about $10 per visitor day and estimates that water is worth $81 million.[15] Although this recalculation quadruples the noncash benefits estimated by Clawson, it fails to bring benefits above costs.

Clawson's results contradict the Forest Service's argument that benefits, including noncash benefits, are greater than costs. George Leonard, the associate chief of the Forest Service, responds by calling Clawson's charge of 5 percent of the asset value "absolutely contrary to the general concept of sustained yield" because it suggests that the national forests could be rapidly cut over to convert the assets into cash.[16] Other agency officials say that Clawson failed to account for some noncash benefits of forest management.

Yet one resource can be shown to lose money without question: timber. Of national forest resources, only timber is sold on the open market. Competitive bidding ensures that most timber sales bring the highest possible value. Yet timber is as inefficiently managed as any other national forest resource. A close examination of the timber program will reveal a great deal about the Forest Service as a whole and why it is inefficient. Before making that examination, however, it is worthwhile to review the agency's organization. While many readers of this book will be familiar with this information, it is included for those who have not previously been involved in national forest issues.

Organization of the Forest Service

The Forest Service divides its operations into three branches: Research, State and Private Forestry (assistance to other forest land owners), and the National Forest System (figure 1.1). The National Forest System represents by far the largest of the triad, consuming over 90 percent of the agency's budget.[17] The national forests are managed by four levels of organization: the Washington office, regional offices, national forests, and ranger districts.

In his headquarters in the monolithic South Agriculture Building in Washington, DC, the chief forester, or chief, directs the Forest Service, along with a staff that includes an associate chief, five deputy chiefs, and numerous associate deputy chiefs, directors, and other officials. The chief is the only director of a federal agency who is a career employee appointed from the ranks rather than a political appointment made by each new administration. The Forest Service is nominally overseen by the politically appointed assistant secretary of agriculture in charge of natural resources and the environment. But the Forest Service's many friends in Congress help to ensure that the administration in power has little effect on the agency.

Beneath the Washington office are nine regional offices, numbered 1 through 10; Region 7 disappeared in a merger in 1966 (figure 1.2). Each regional office is guided by a regional forester, who is assisted by deputies, directors, and other officials.

Each region oversees two to nineteen administrative units, often called *national forests*. The term is used with caution as there are 156 different national forests but

Figure 1.1 Forest Service Organization Chart

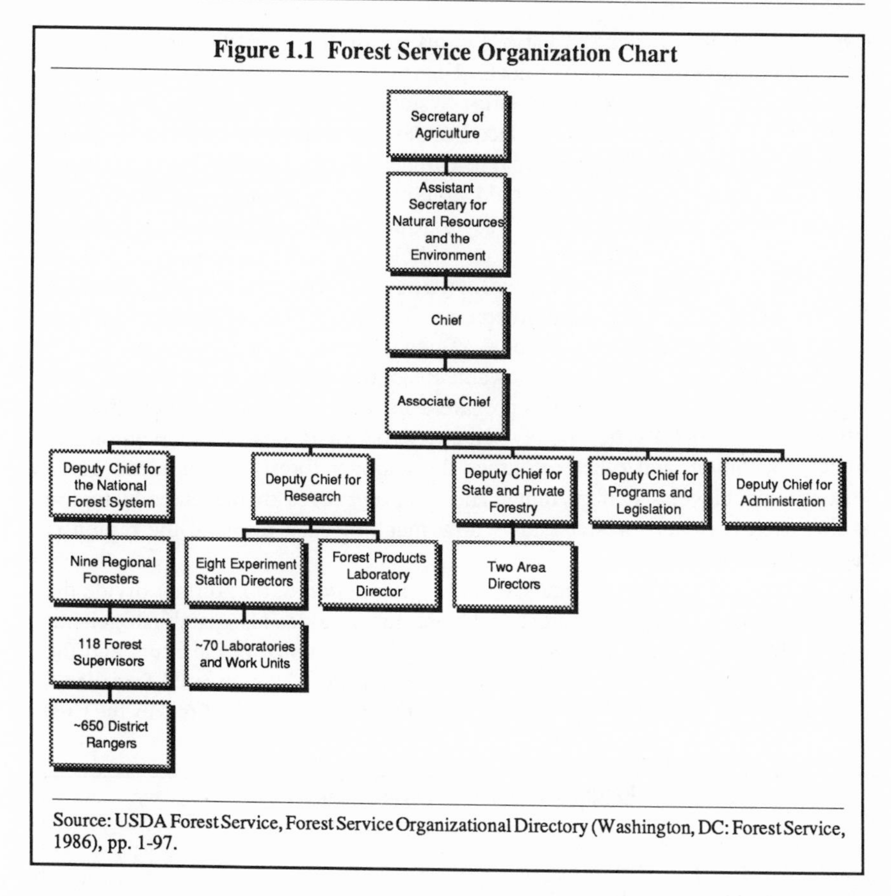

Source: USDA Forest Service, Forest Service Organizational Directory (Washington, DC: Forest Service, 1986), pp. 1-97.

only 119 administrative units. National forests are proclaimed only by Congress, but for administrative convenience, the Forest Service often manages two to six forests as a single unit. The administrative units are each managed by a forest supervisor, who is assisted by a deputy supervisor and other managers.

The lowest level of the National Forest System is the ranger district. Each forest supervisor oversees about six district rangers, for a total of about 650 ranger districts. A few forests are so small that they have no districts, while two forests have twelve or more districts. The ranger districts carry out all the on-the-ground work, including preparing and administering timber sales, reforestation, maintenance of recreation facilities, and range management.

Collectively, the chief, nine regional foresters, 118 forest supervisors, and 650 district rangers are called *line officers*. All other national forest officials are *staff officers*. The Research and State and Private Forestry branches of the Forest Service have similar lines of authority, but on a smaller scale.

Research, for example, is divided into eight experiment stations and a forest products laboratory, each headed by a station director. Most of the stations have several laboratories located in various cities within their jurisdictions, and each laboratory divides its work into projects. Note that the only connection between the Research branch and the National Forest System is at the level of the chief. Early Forest Service researchers were placed under the authority of National Forest System managers, with the result that researchers sometimes found their research suppressed or distorted so that it would support "official policy." To prevent this from happening, research was made a separate branch answerable only to the chief.

The Forest Service is one of the most publicly accessible agencies in the federal government. Virtually all management activities are carried out according to a plan that was prepared with public involvement. The current planning process, like the agency itself, has four levels: national, regional, forest, and project plans.

The Forest and Rangelands Renewable Resources Planning Act (RPA) of 1974 requires the Washington office to prepare a national plan for all Forest Service activities.[18] This *RPA program*, revised every five years, describes the outputs of timber, recreation, and other resources the national forests can produce given various budget levels. Timber, range, and other *objectives* are presented by region so that each regional forester knows how much of each resource the region is expected to produce.

Regional offices prepare plans, called *regional guides*, that further divide the objectives for each national forest.[19] The regional guides also establish general policies for the national forests such as the maximum size of clearcuts and the minimum qualifications land must meet to be considered commercial forestland. Although the regional guides are prepared by the regions, the chief makes the final

Figure 1.2 Forest Service Regions

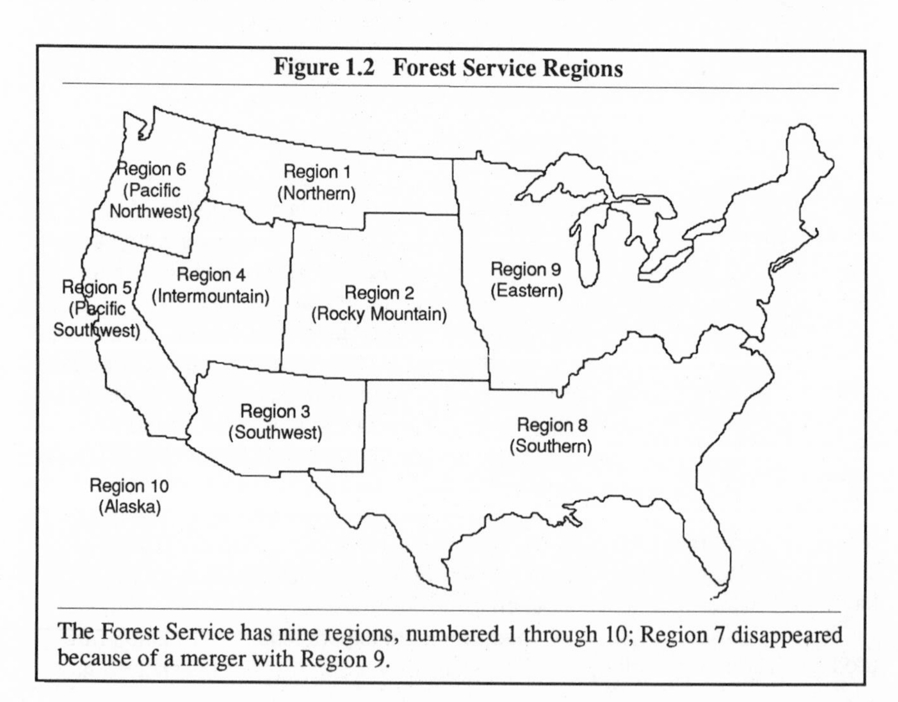

The Forest Service has nine regions, numbered 1 through 10; Region 7 disappeared because of a merger with Region 9.

decision on all questions. Regional guides were first completed in the early 1980s and are to be revised every ten to fifteen years.

Forest plans prepared by the national forests state exactly how each acre of forestland is to be managed and how much timber and other resources will be extracted each year. Although forest planning began in 1979, most national forests had not yet finished their plans by the end of 1986 and thus were operating under older plans. Final decisions on all plans are made by the regional foresters. Plans are to be revised every ten to fifteen years.

All of the plans just described— the RPA program, regional guides, and forest plans — are prepared in draft and final formats, with the public invited to comment on the draft plans and at several other stages of the planning process. Project plans, the final level of plans, may occasionally have draft and final formats but usually do not involve as much public comment as the others. Individual project plans are prepared by the ranger districts for all major activities, such as timber sales, road construction, and construction of recreation facilities.

As described here, Forest Service planning appears to be from the top down, with RPA objectives determining the levels of timber and other outputs that will be produced by the national forests. However, information actually travels in both directions: The RPA program passes information about national demand for forest resources to the forest plans, while the forest plans pass information about the individual forests' capabilities of meeting those demands back to RPA planners. Thus, the results of forest plans have a strong influence on subsequent RPA plans.

A similar process is used for budgeting. Local forest officials begin to estimate their budgetary needs several years in advance. Budgets are compiled by higher

Figure 1.3 Forest Service Budgets Compared with Requests

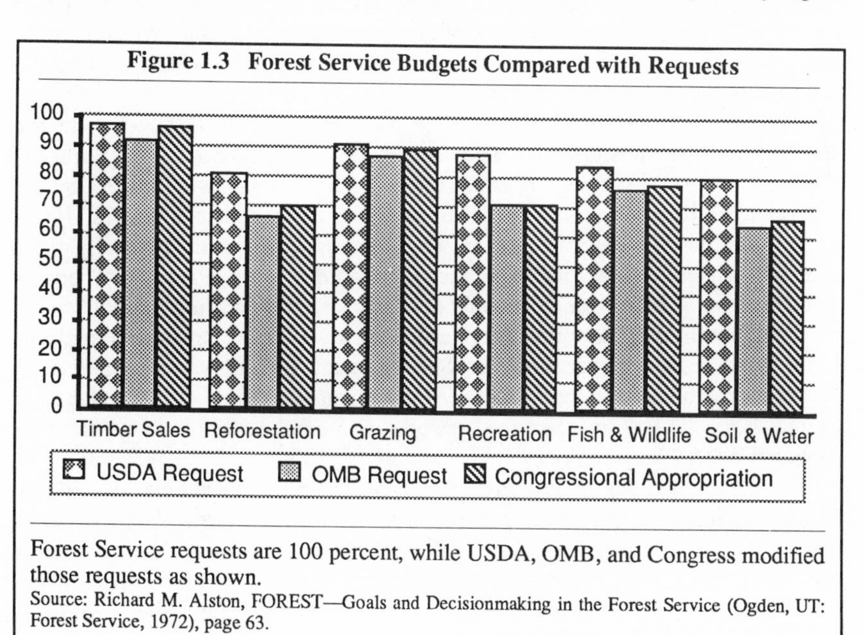

Forest Service requests are 100 percent, while USDA, OMB, and Congress modified those requests as shown.

Source: Richard M. Alston, FOREST—Goals and Decisionmaking in the Forest Service (Ogden, UT: Forest Service, 1972), page 63.

officials up the line of command until they reach the Washington office. The Washington office's budget is reviewed and usually altered by the secretary of agriculture and altered more drastically by the Office of Management and Budget (OMB). The final budget is approved by Congress.

In 1972, Richard Alston reviewed eighteen years of budgetary history and found that the Department of Agriculture and OMB each made large reductions in proposed Forest Service budgets for most resource areas. However, the reductions for timber sales were relatively small. Although Congress tended to increase most resource budgets above the levels proposed by the OMB, timber and range were the only resources to receive 90 percent or more of the funds originally requested by the Forest Service (figure 1.3). Fish and wildlife, recreation, soil and water, and research received only 60 to 80 percent of the original Forest Service requests.[20]

National forest managers work under the guidance of the *Forest Service Manual* (often abbreviated *FSM*), a 25,000-page work that occupies as much as seven feet of shelf space in most forest offices. The basic manual is frequently supplemented by the Washington office, regional offices, and forest supervisors offices. Managers are given additional direction through memoranda discussing specific topics.

A Brief History

The national forests were first conceived in 1891, when a law was passed allowing the president to designate public lands as "forest reserves." The 1897 Organic Administration Act allowed timber sales and other management activities in these reserves and specified that the goals of the reserves were to "improve and protect the forests . . . , securing favorable conditions of water flows, and to furnish a continuous supply of timber."

The Forest Service itself did not come into existence until 1905, when the forest reserves were transferred from the Department of the Interior to the Bureau of Forestry in the Department of Agriculture. That bureau was headed by the charismatic Gifford Pinchot, who changed its name to Forest Service and gave himself the title of chief forester. Although nominally under the authority of the secretary of agriculture, Pinchot in fact reported directly to President Theodore Roosevelt and was considered a member of Roosevelt's "kitchen cabinet."

Traditionally, the secretary of agriculture has always been more concerned with farms than forests, and thus the Forest Service has always enjoyed more autonomy than most bureaus. Even today, most people and many agency officials refer to the agency as the "U.S. Forest Service," despite an internal campaign to use the more proper "USDA Forest Service."

By 1891, when the act authorizing creation of forest reserves was passed, few lands remained in federal ownership in the eastern United States. A law passed in 1911 authorized the Forest Service to purchase forested or cut-over lands in the East for "the regulation of the flow of navigable streams or for the production of timber." Under this law, national forests are now scattered throughout the South, Midwest, and Northeast, but most are still concentrated in the West.

With plenty of timber still available on private lands, the demand for national forest timber was low through the early 1940s. National forest management up to that time was largely custodial in nature, oriented toward forest protection, research, and administration of grazing, recreation, and other forest uses.

The concept of designating an area as "wilderness," withdrawing it from timber or other developments, was first conceived by a Forest Service landscape architect named Arthur Carhart.[21] The idea was strongly promoted by two other Forest Service employees, Aldo Leopold and Robert Marshall. Leopold went on to found the wildlife management profession, and together with Marshall, cofounded the Wilderness Society. Marshall became the Forest Service's director of recreation and convinced the agency to classify over 5 million acres of land as wilderness or primitive areas.

The post-World War II housing boom, combined with the depletion of private timber supplies in many parts of the country, increased demand for national forest timber. To gain access to this timber, the Forest Service embarked on a large-scale road construction program. The volume of timber national forest managers were allowed to sell increased from 5.6 billion board feet in 1950 to 12.8 billion by 1968.[22]

In many cases, agency managers found that some of the best stands of timber were within the boundaries of classified wilderness areas. Since wilderness was an administrative, rather than legal, classification, the Forest Service could simply change the boundaries and build roads into the former wilderness areas. This led to increasing controversy, and in 1956 the Wilderness Society and other conservation groups began to ask Congress for legal protection of wilderness areas.

Such protection was not immediately forthcoming, largely due to opposition from the Forest Service. Instead, the agency offered its own new legislation, the Multiple-Use Sustained-Yield (MUSY) Act. Passed in 1960, this law required the Forest Service to manage the national forests for "outdoor recreation, range, timber, watershed, and wildlife and fish purposes." Wilderness supporters in Congress added a statement to the Forest Service's proposed language, which said that wilderness was "consistent with the purposes and provisions of this Act."

While many analysts believe that MUSY was a sincere attempt by the Forest Service to ensure that the national forests were not predominantly managed for timber, in reality the act contains little concrete guidance for the forest manager. Indeed, the U.S. Court of Appeals for the Ninth Circuit once reviewed the act and concluded that it "breathes discretion at every pore."[23]

Although wilderness supporters were able to convince Congress to pass the Wilderness Act in 1964, increasing timber sale levels caused continuing controversies over the appropriate management of lands outside the wilderness areas. These controversies were fueled by the increasing use of clearcutting, with its dramatic visual impact. Before 1960, most national forests used selection cutting, which does not have a strong visual impact, but by 1970 nearly all national forests used clearcutting as the predominant harvest method.

Controversies were particularly strong in Montana and West Virginia.[24] In Montana's Bitterroot National Forest, the Forest Service used bulldozers to terrace hillsides to ease reforestation after clearcutting. However, a 1970 report prepared by the dean of the University of Montana School of Forestry and several university professors found that such practices "cannot be justified as an investment for producing timber."[25] Moreover, the report, which was prepared at the request of the Senate Interior and Insular Affairs Committee, concluded that "multiple use management, in fact, does not exist as the governing principle on the Bitterroot National Forest."[26]

The Forest Service responded to these controversies by developing a detailed process for planning the future management of the national forests. Timber management plans would be prepared for each forest to determine how much timber could be cut each year. Land use plans would be prepared for "planning units" — which averaged about one-fifth of a forest — to determine which lands would be managed for timber and which for other resources. Both sets of plans would involve heavy public participation and would be revised every ten years.

A factor speeding the planning process was a court decision that the Forest Service could not build roads and sell timber in roadless areas — areas that were potentially suitable for congressional designation as wilderness — until it had completed an environmental impact statement (EIS) weighing wilderness values against timber values. Twice the Forest Service attempted to prepare a nationwide *roadless area review and evaluation* (RARE), but courts ruled that local analyses were required to seriously consider all the factors affecting a decision to develop a roadless area.

While the wilderness versus timber debate was fought in the unit plans, the timber management plans were the scene of a debate over the appropriate level of national forest timber harvests. Even before passage of the Multiple-Use Sustained-Yield Act, the Forest Service had a policy of selling timber on a sustained-yield basis, meaning that no more timber would be sold at any time than could be sold in the future. In 1946, for example, the regional forester for Oregon and Washington forests, H. J. Andrews, defined sustained yield as "the greatest even flow of forest products in perpetuity which the forest lands involved are capable of producing."[27] He added that "it is and has been the policy of the Forest Service since its inception over forty years ago to limit their harvesting from National Forest management units to their sustained yield capacity."[28]

In 1969, however, the Forest Service published a disturbing report called the *Douglas-Fir Supply Study*. This report found that forests in western Oregon and western Washington would experience an allowable cut *falldown* of as much as 30 percent when existing stocks of old-growth were depleted.[29] Such a falldown clearly violated Andrews's definition of sustained yield.

The timber industry argued that such a definition was too strict. Led by Carl Newport, of Portland's Mason, Bruce, and Girard forestry consulting firm, and Wesley Rickard, a Washington forestry consultant, industry representatives pointed out that such an *even-flow* policy wasted timber that could otherwise be profitably used.[30] But in 1973 the Forest Service directed national forests to observe a strict *nondeclining flow* policy in future timber management plans.[31] The Forest Service believed that falldowns could be minimized if funding were available to intensively manage second-growth forests. Congress passed RPA in 1974 to help provide such funding. Under the act, the Forest Service would prepare an RPA program every five years to indicate how much timber and other resources could be produced given proposed funding levels. The first RPA program was prepared in 1975.

Meanwhile, West Virginia clearcutting controversies were coming to a head. While working for the Natural Resources Defense Council, Lawrence Rockefeller, great-grandson of the oil billionaire, found an obscure portion of the 1897 Organic Administration Act that seemed to forbid clearcutting. Opponents of clearcutting used this law to halt cutting in the Monongahela National Forest. An appeal by the Forest Service confirmed this interpretation.

Forest Service officials and timber industry representatives feared that national forest timber sales would fall dramatically if the law were enforced nationwide. Environmentalists countered that clearcutting was symptomatic of many problems with the agency. The debate led to passage of the National Forest Management Act (NFMA) of 1976. NFMA was supposed to resolve controversies, but it left many questions open for further dispute. While not forbidding clearcutting, it limited the circumstances under which the practice could be used by the Forest Service. While not requiring nondeclining flow, it allowed the Forest Service to depart from nondeclining flow only for multiple-use purposes.

NFMA affirmed the Forest Service's planning process and established detailed requirements the plans must meet. The Forest Service responded by redesigning its planning process. Instead of separate plans for timber and land use, one plan would cover all resources and land allocations for an entire national forest. The new forest plans would also be closely tied to the RPA program. The program established timber, range, and other objectives for each of the nine Forest Service regions. Regional plans (later called regional guides) would assign the objectives to each national forest. At least one alternative in the environmental impact statement accompanying each forest plan would attempt to meet all RPA objectives.

Forest Service officials were quick to point out that passing objectives from RPA to forest plans did not mean the planning process was "top down." Forests that could not meet their RPA objectives would pass this information up through the regions to RPA planners, who would use it to make the next RPA program more realistic. In fact, this scenario has often occurred. Yet many planners confess that they are under pressure to meet the RPA targets, and some plans have been rejected by higher officials because they did not meet their objectives.

The regional RPA targets were less flexible than those of the forests. Memos from the Washington office state that the regions were to "meet or exceed RPA targets." If one forest plan was unable to meet its timber target, the regional forester would be required to increase the target for another forest.[32]

Congress suggested that NFMA plans be completed by September 30, 1985. In 1976, nine years seemed to be plenty of time to prepare plans that were to last for ten years. But the planning process did not officially begin until 1979, when the Forest Service published regulations governing the planning process.

In 1982, the Reagan administration published new rules, ostensibly to "streamline" the planning process, that in reality delayed many plans as planners worked to meet the new requirements. In 1983, unsatisfied with the Forest Service's response to the new rules, the Reagan administration threatened to replace the chief or regional foresters with political appointments unless plans improved. As a result, all plans in Oregon and Washington were started over from scratch. Many other plans, including those for the Nantahala-Pisgah (NC), Bridger-Teton (WY), Shasta-Trinity (CA), and Boise (ID) forests, were also delayed by administration direction. By the September 1985 deadline, fewer than half the projected 123 plans had been published in final form. A year later, plans for more than 20 national forests had not yet even reached the draft stage.

Planning is also very expensive, partly because of delays. A General Accounting Office study of two Idaho plans estimates that well over $2 million have been spent to date on each — and the plans are still incomplete.[33] The total costs of the 123 forest plans will probably exceed $300 million.

Despite the time and money invested in planning, few people are satisfied with the results. Although the timber industry has worked hard to promote increases in timber sales, only four forests have seriously considered alternatives to nondeclining flow. Although environmental groups have worked hard to eliminate clearcutting, only one forest where clearcutting is commonly used has seriously considered the alternative of selection cutting.

If nothing else, however, the plans have provided the public with a gigantic database. Most of the plans provide detailed information about the costs and benefits of timber, grazing, recreation, wildlife, and watershed management. This information has been supplemented by public reviews of the plans. CHEC, a forestry consulting firm, has reviewed more than forty forest plans for state agencies and citizens' groups. These reviews, which include detailed analyses of Forest Service computer models, timber sale records, and other information, provide additional insights into the Forest Service.

The following four chapters draw upon this information to show that most national forests are selling more timber than can be economically or environmentally justified. Chapter 2 shows that timber sales on most national forests lose money. Chapter 3 shows that investments in reforestation and other timber management practices are not worthwhile in any but a few national forests. Chapter 4 examines the Forest Service's planning process to demonstrate that plans are biased toward timber even though timber loses so much money. Chapter 5 responds to Forest Service claims that money-losing timber sales are needed to provide or protect other resources such as recreation and wildlife.

Notes

1. USDA Forest Service, *1987 Budget Explanatory Notes* (Washington, DC: Forest Service, 1986), p. 2.
2. Ibid., pp. 11–12, 150.
3. Ibid., pp. 183–184.
4. Ibid., pp. 164–165.
5. Ibid., pp. 171–172.
6. Ibid., p. 189.
7. "The Fortune 500," *Fortune*, 28 April 1986, pp. 372–388.

8. USDA Forest Service, *A Recommended Renewable Resources Program: 1985-2030* (Washington, DC: Forest Service, 1986), p. 12.
9. USDA Forest Service, *1987 Budget,* pp. 11–12.
10. Ibid., p. 12.
11. Marion Clawson, *The Economics of National Forest Management* (Baltimore, MD: Johns Hopkins University Press, 1976), p. 56.
12. USDA Forest Service, *Renewable Resources Program*, p. 12.
13. "The Role of 'Below Cost' Timber Sales in National Forest Management," memo on file at the Forest Service, Washington, DC, 16 August 1984, p. 1.
14. Clawson, *Economics of National Forest Management*, p. 56.
15. USDA Forest Service, *1985-2030 Resources Planning Act Final Environmental Impact Statement (Washington*, DC: Forest Service, 1986), pp. F-5–F-10.
16. Randal O'Toole, "Interview With the New Chief and Associate Chief of the Forest Service," *Forest Watch* 7(11):15–20.
17. USDA Forest Service, *1987 Budget*, p. 12.
18. Forest and Rangelands Renewable Resources Planning Act of 1974, section 4 (16 USC 1602).
19. 36 *Code of Federal Regulations* 219.4.
20. Richard M. Alston, *FOREST — Goals and Decisionmaking in the Forest Service* (Ogden, UT: Forest Service, 1972), p. 63.
21. John V. Krutilla, *Resource Availability, Environmental Constraints, and the Education of a Forester* (Washington, DC: Resources for the Future, 1977), pp. 27–28.
22. Michael Frome, *The Forest Service* (Boulder, CO: Westview Press, 1984), p. 111.
23. Perkins v. Bergland, 608 F. 2d 803, 806 (9th Cir. 1979).
24. Daniel R. Barney, *The Last Stand: Ralph Nader's Study Group Report on the National Forests* (New York, NY: Grossman, 1974), pp. 5–6, 41–42.
25. Arnold W. Bolle et al., *A University View of the Forest Service* (Washington, DC: U.S. Senate Committee on Interior and Insular Affairs, 1970), p. 13.
26. Ibid.
27. "Hearing Record for the Proposed Shelton Cooperative Sustained Yield Unit," on file at the Olympic Forest supervisor's office, Olympia, WA, 18 September 1946, pp. 10–11.
28. Ibid., p. 11.
29. USDA Forest Service, *Douglas-Fir Supply Study* (Portland, OR: Forest Service, 1969), p. 14.
30. Letter from Carl Newport to Western Forest Industries Association reviewing Gifford Pinchot National Forest Timber Management Plan, 7 June 1974, 7 pp.
31. USDA Forest Service, "Emergency Directive No. 16," memo from chief to regional foresters, May 1973, 3 pp.
32. "RPA Targets and National Forest Land and Resource Management Planning," memo from the chief to regional forester, R-6, 14 September 1981, on file at Portland, OR, office of Forest Service, 1 p.
33. General Accounting Office, *Forest Planning Costs at the Boise and Clearwater National Forests in Idaho* (Washington, DC: GAO, 1986), pp. 1–2.

Chapter Two
Timber Sales Below-Cost

The Forest Service has long been respected as one of the few "efficient" agencies in government. In 1981, for example, Pennsylvania State University and the U.S. Office of Personnel Management called the Forest Service an "excellent organization."[1] Imagine that Lee Iacocca has hired a former chief of the Forest Service as a consultant to tell Chrysler Corporation how it can apply Forest Service principles so that Chrysler, too, can be an excellent organization. After several months of study, the chief presents a report to Chrysler's board of directors.

"I propose that a new system be used for selling automobiles," announces the chief. "Chrysler should sell all its cars at weekly auctions held by the dealers. That way each car will be guaranteed to sell for the highest price anyone would be willing to pay for the car."

"What if no one is willing to pay very much for a particular car?" asks a board member.

"That's one of the advantages of this system — almost all cars will be sold for some price, so Chrysler won't have to worry about whether people prefer one kind of car over another. Of course, no car will be sold for less than a minimum price equal to $100 plus the cost of transporting a replacement car from the factory to the dealer."

"That could be a very low price," comments a board member.

"Yes, but remember that the replacement car itself will be sold, and it may bring a higher price. If the first car were not sold, there would be no room for the replacement and no chance of obtaining that higher price."

"What about trucks?" asks a board member. "We sell many of our trucks in fleets to large companies, not one at a time."

"Trucks would be priced by a special formula," answers the chief. "The formula subtracts the costs of operating a truck from the estimated value of the truck to an average truck owner. The amount that is left over is called the residual price. Using this formula, Chrysler could easily outsell GMC and other truck manufacturers."

"What evidence do you have that Chrysler can make money from this program?" asks a skeptical board member.

"Your economists ran cost and revenue figures into a Forest Service computer program, which concluded that the Plymouth and Dodge divisions would lose money on almost every car they sell. But these losses would be more than made up for by the expensive luxury cars sold by the Chrysler division."

"If that's true," notes a board member, "why don't we just sell Chryslers? That would increase our profits even more."

"If you did that, someone else — perhaps the Brazilians or Canadians — would meet the demand for automobiles," answers the chief. "It's important for Chrysler that all of these cars be locally produced."

"I'm looking at your report," says a board member, "and the capital investments you propose to build new factories aren't all included in your benefit-cost analyses."

"Many of the capital costs were ignored because they would be paid for by profits on sales of cars last year," answers the chief. "We amortized the other capital costs over fifty years because we assumed that they would eventually be paid for by the sale of automobiles."

"So, to sum up," says Iacocca, "you are telling us that most of the cars we sell will lose money, but overall we will make money provided we don't count capital costs. Is that right?"

"That's exactly right," says the chief. "That's the way the Forest Service has operated for over forty years."

Chrysler will probably not follow the chief's recommendations even though this is an almost perfect representation of Forest Service timber sale and accounting practices. Forest Service timber is sold in competitive auctions, which the Forest Service says proves that it is sold for its "fair market value."[2] But the minimum bid price is often no more than $0.50 per thousand board feet plus the cost of reforesting the land after the timber is cut. The Forest Service says that losses from such sales are justified by the fact that the reforested trees might be sold at a profit in the future.[3]

The minimum bid price for other timber is a residual price equal to the value of the timber to the purchaser minus the cost of logging and manufacturing the timber. This is still often much less than the cost of sales, particularly in Alaska, the Rocky Mountains, and the Northeast. The Forest Service excuses this by saying that these losses are more than made up for by profits in the Pacific Northwest, and that "as a whole, the timber sale program makes money."[4]

A close examination reveals that, as a whole, the timber program often does *not* make money when interest charges and capital costs are considered. The Forest Service ignores realistic interest charges, saying that they are "absolutely contrary to the concept of sustained yield."[5] Capital costs are sometimes amortized over lengthy periods but are often completely ignored.[6] Given all these practices, it is not surprising that one Forest Service economist recently noted: "Timber programs are not inherently evil. Neither is it evil to make money. Somehow, though, public foresters tend to believe that it *is* inherently evil to have a timber program that makes money."[7]

Any private company using these business practices would quickly go bankrupt. The Forest Service can continue to practice them only because it has a sympathetic "banker" — the U.S. Congress — which jumps at the thought of jobs in each of the forty states in which there are national forests. Yet the Forest Service timber program is a business, one that competes with Weyerhaeuser, International Paper, and thousands of other private timber land owners. Shouldn't Forest Service business practices be the same as those of private owners? A number of studies made in the past fifteen years have shown that the use of standard business practices would save taxpayers hundreds of millions of dollars each year and promote management of timber on private lands, which are potentially more productive than the national forests.

Most of the studies fall into two major classes: *cash flow* and *bare land value*. Cash flow studies are designed to estimate the profits or losses from sales of timber now standing in the forest. Bare land value studies, discussed in chapter 3, estimate the return on investments in reforestation and other practices designed to grow new stands of trees. Together, these studies provide a reasonable picture of the economic potential of national forest timber management.

Cash Flow Studies of Timber Sales

Cash flow analysis is the most popular method of reviewing timber sale economics because budget and receipts data are easy to collect. However, it can be deceptive because money spent in one year is not necessarily related to the receipts for that same year. For example, many cash flow studies measure reforestation costs against current timber receipts, yet reforestation is really an investment in future timber growth. Most cash flow studies also count at least some road construction costs, even though roads may be used to provide access to future sales.

In addition, timber sale receipts are subject to market fluctuations. The Forest Service attempts to sell about the same amount of timber each year, so expenditures remain relatively constant. But purchasers are allowed three to five years to harvest the timber they buy, and until recently they were not required to pay for timber until they cut it. Thus, receipts go down during poor wood markets and up during good markets. A sound cash flow study must include some good years and some poor years.

In the first major cash flow study, Tom Barlow, of the Natural Resources Defense Council (NRDC), compared timber receipts with timber management costs between 1970 and 1973 for each of the nine Forest Service regions. The 1975 report counted as "costs" line items for timber sales — including *sale preparation, harvest administration, silvicultural examinations* (silviculture is the science of timber management), and *timber resource planning* — and timber roads, including new *road construction* and *road engineering and design* but not timber purchaser roads. Barlow found that most regions routinely spent more money on timber than they collected, and all but Regions 5 (California), 6 (Oregon and Washington), and 8 (the South) spent more than they collected at least some of the time. Barlow noted that the Forest Service timber program as a whole "shows a net profit, [but] a proper cost accounting system would show that many individual sales are made at net losses."[8]

In 1977, William Hyde, then with Resources for the Future, focused on a 1976 timber management plan adopted for Colorado's San Juan National Forest, which proposed to harvest timber from 169,000 acres of roadless land. Using Forest Service data, Hyde found that timber sales returning less than $38.70 per thousand board feet would lose money. Yet the average high bid price in 1976 was only $2.65 per thousand board feet. The average annual sale price had been less than $24 since at least 1970 and had exceeded $7 in only two out of the seven years. Occasional sales did receive bids as high as $97 per thousand board feet, but they were in easily accessible areas. Hyde concluded that sales in roadless areas would rarely, if ever, justify their costs.[9]

In 1980, NRDC fine-tuned Barlow's earlier study by comparing receipts with costs for each individual national forest. The study covered the years 1974 through 1978, a period when timber prices were relatively high. Against timber receipts, NRDC counted reforestation, timber stand improvement, and certain other management costs, along with timber sale preparation and administration costs. Because of doubt about whether roads served just timber or provided multiple-use benefits, NRDC researchers decided to make one comparison using all timber road costs and a second comparison using only half of the timber road costs.

The data, published in a legal-sized book titled *Giving Away the National Forests*, were startling. When all road costs were counted against timber receipts, 73 of the 118 national forests spent more on timber over the five-year period than they collected. Even when only half of the road costs were counted, 66 forests lost money.[10]

Gloria Helfand, an NRDC economist who did much of the data collection, later analyzed the results in more detail and found that 63 forests lost money between 1974 and 1978 even when no road costs were counted. In fact, if only the sale preparation and administration costs were charged against timber, 46 forests still

Figure 2.1 National Forests Losing Money on Timber, 1974-1978

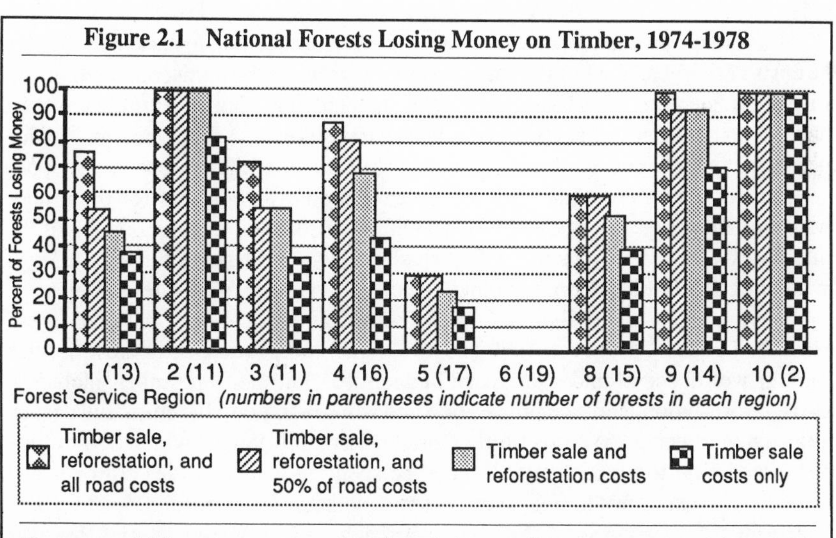

No matter which costs are counted against timber receipts, many national forests fail to produce receipts equalling those costs.

Source: Gloria E. Helfand, "An Analysis of the Costs and Receipts of National Forest Timber Sales," masters thesis on file at Washington University, St. Louis, MO, p. 102.

spent more than they returned to the Treasury (figure 2.1).[11]

The Forest Service countered NRDC's results by saying that it was inappropriate to count road costs against timber since roads provided so many multiple-use benefits. In addition, the agency belittled the NRDC study by saying that only 22 percent of the volume of timber sold by the Forest Service lost money. In 1984, the Forest Service claimed that only a third of the timber sold in 1983 — which, it pointed out, was a very poor year for the timber industry — lost money.[12]

These points only obscured the issue. As Helfand found, NRDC's data clearly indicated that most forests lost money even when roads were not counted. The Forest Service never revealed how it arrived at 22 percent, and the figure is probably an underestimate. Helfand admitted that the 72 forests which NRDC indicated lost money actually produced only 20 to 25 percent of the volume of national forest timber sales. Yet those forests were managing timber on tens of millions of acres. After all, six to fifteen acres of timber must be harvested from a money-losing forest such as the Bridger-Teton or Gallatin to produce the same volume as one acre in a profitable forest such as the Siuslaw.[13] In turn, the environmental effects of timber harvests on a money-losing forest were considerably larger than those from the harvest of an equal volume elsewhere.

The Wilderness Society prepared an "issue brief" in 1984 comparing high bids for timber sales in a given year against timber-related costs during that year. The costs included timber sale preparation and administration, costs of mitigating the effects of timber on other resources, road construction and engineering costs, and reforestation and timber stand improvement costs paid out of the Knutson-Vandenberg reforestation fund. Timber's share of general administration costs were also added.[14]

To avoid the problem of counting receipts from timber sold in past years against the costs of this year's sales, the study counted as receipts the undiscounted high bid price of timber. The study found that 97 forests spent more money on timber in 1982 than they expected to collect from sales that year. Of these, 75 spent more than twice their expected receipts, while only 4 spent less than half of their expected receipts.[15]

The study admitted that 1982 was an exceptionally poor year for the timber industry. However, the study compared similar costs and receipts for each of the nine Forest Service regions in 1978, a much better year. Five of the nine regions spent over $62 million more than they expected to collect from sales that year.

In 1983, a Library of Congress study focused attention on one national forest, Alaska's Tongass. Between 1970 and 1984, the report found, the Forest Service spent or would spend under proposed budgets $316 million on timber, including roads, reforestation, and sale preparation and administration. In return, the agency collected or expected to collect little more than $62 million, of which $16 million was turned over to the state of Alaska to be spent by local counties on roads and schools. The net return to U.S. taxpapers was negative $313 million.[16] At no time during the 15-year period did income approach expenses, and the net loss was increasing each year.

In early 1984, the Library of Congress prepared a new report reviewing national forest timber budgets on a state-by-state basis. The report covered an 11-year period between 1973 and 1983 and found that in only 16 of 39 states with timber-producing national forests did the revenues exceed costs in at least 6 of the 11 years. These states were mostly on the Pacific Coast and in the South.[17] In most of the remaining states, the study said, "a negative cash flow occurred in every year" of the study. In fact, in 21 states the revenues were less than one-third of the costs of laying out and administering timber sales (including silvicultural examination and timber resource planning). This included most of the Rocky Mountain states, Alaska, and states in the Appalachian mountains..

Many people use the term *deficit sales* to refer to sales that cost the Forest Service more than they returned. However, the Forest Service applies the term *deficit sales* to sales that appraisers expect will cost the purchaser more than the timber is worth. To prevent confusion, NRDC coined the term *below-cost timber sales* or simply *sales below cost.* Forests that, according to the NRDC or Library of Congress studies, frequently lost money were called *sales-below-cost forests.*

Even this terminology was deceptive, however, for it implied that all timber in the below-cost forests lost money. More subtly, it implied that all timber sold by the remaining forests made money. In fact, neither NRDC nor the Library of Congress examined any timber sales. They merely looked at the overall budgets and receipts of entire national forests or, in the case of the Library of Congress, all the national forests in each state.

In fact, some of the timber sold by some of the below-cost forests probably made money. This net income would tend to reduce the total net losses calculated when comparing budgets and collections, so the losses experienced by many of these forests were underestimated. Also, some of the timber sold by the above-cost forests lost money. The losses from these timber sales are obscured by the money made from selling other timber in those same forests. Thus, the NRDC and Library of Congress reports greatly underestimated the total losses to taxpayers caused by below-cost timber sales.

To partially rectify this situation, the General Accounting Office (GAO) issued a report in June 1984 that examined timber sales in four Forest Service regions. In addition to looking at sales in below-cost regions like Regions 1 (Montana and northern Idaho), 2 (Colorado, eastern Wyoming, South Dakota), and 3 (Utah, Nevada, southern Idaho), the GAO examined sales in the above-cost Region 6 (Oregon and Washington).

The GAO compared sale preparation and administration and road costs to the high bids for 3,244 sales sold in 1981 and 1982. Over 90 percent of the sales examined in Regions 2 and 4 lost money, while 60 percent lost money in Region 1, and 24 percent lost money in Region 6.[18] The losses from these sales totaled $156 million over the two-year period. Although receipts from the remaining sales more than covered the losses, the government would have been at least $156 million richer had the money-losing sales not been made.[19]

The Forest Service Responds

With critiques from the Library of Congress and GAO in hand, the Interior Subcommittee of the House Appropriations Committee began to closely review Forest

Service timber sales. In response, the Forest Service decided to make a detailed reply to charges that it was wasting taxpayers' money. In August 1984, the agency distributed a white paper called "The role of 'below cost' timber sales in national forest management."

As might be predicted from the title, the Forest Service contended that timber sales, including money-losing sales, were a necessary part of national forest management. "Maximizing monetary profit is not the primary objective of National Forest management," said the paper. Instead, "by law, these public forests are managed for numerous values other than the money they can produce."[20]

The paper claimed that timber sales were often "the most effective manner" of investing in "long-term future timber growth" and "improv[ing] wildlife habitat sites or watersheds." The paper called roads "another benefit of timber sales," saying that they "provide access to later harvests," "make possible protection of the forests from fire," and "enhance recreation, wildlife management, grazing, and other uses of the forest."[21] These claims will be discussed in greater detail in chapter 5.

After listing these *side benefits* to timber management, the white paper went on to point out that "year after year, the value of timber sold from the national forests exceeds timber sale costs." Even though most forests lost money, the ones that made money more than made up for it. Thus, "by any measurable standards, the American taxpayer is getting a fair return on his [*sic*] investment in national forest timber management and is not subsidizing the timber industry through appropriations."[22]

The Forest Service is arguing that money-losing sales in some national forests are justified by the fact that sales in other national forests, which may be a thousand miles away, make enough money to cover the losses. This is analogous to the chief telling Lee Iacocca that it is all right for Plymouth and Dodge divisions to lose money so long as the Chrysler division makes up the difference. A company that tolerates such attitudes makes an easy takeover target: Money-losing divisions would reduce the total asset value of the company, so an outsider could buy the company for a low price, clear out the inefficient managers, sell off the unpromising divisions, and end up with a company whose value is much higher than the purchase price.

The Forest Service white paper went on to argue that below-cost sales in particular are not a subsidy to the timber industry because "National Forest timber is sold competitively for its fair market values."[23] This statement implies that the Forest Service appraisal process correctly assesses fair market value. The law requires the Forest Service to sell timber for "not less than appraised value." In affirming this requirement in 1976, a Senate report noted that it was intended to ensure "that the United States will obtain fair market value for timber and forest products."[24] The Forest Service had always interpreted "appraised value" to mean "fair market value," and, in fact, Forest Service regulations note that "the objective of national forest timber sale appraisal is to estimate fair market value."[25] The *Forest Service Manual* says that "appraised value and fair market value as used by the Forest Service mean the same."[26]

The same part of the manual defines fair market value as "the price acceptable to a willing buyer and seller, both with knowledge of the relevant facts and not under compulsion to deal." However, the manual directs timber sale officers to use an appraisal method called the "residual pricing system."[27]

This system begins with published prices that mills are paying for logs and subtracts the costs of cutting and transporting trees to the mill. The remainder becomes the "indicated advertised rate." The system ensures that the price will be low enough that a timber company can make a profit from buying the timber, but it makes no attempt to ensure that the government will make a profit or at least break even on the sale.

Thus, in claiming that it sells timber for "fair market value," the Forest Service implies that it is selling the timber for a price a private landowner would accept. If this were true, then timber sales could not be considered a subsidy to the timber industry. But no reasonable seller would perpetually agree to sell goods at a loss, so in fact Forest Service appraisal prices often are below the true fair market value.

Despite this white paper, the House Appropriations Committee voted to reduce Forest Service timber sales by 700 million board feet and directed the Forest Service to make the largest reductions in forests whose timber programs had a negative cash flow. The Forest Service estimated the cash flow of each national forest and projected which forests would suffer the greatest reductions in sale levels under the House proposal. The largest reductions would have been in Idaho, home of Senator James McClure, who then chaired the Senate subcommittee in charge of Forest Service appropriations.[28] McClure convinced a Senate-House conference committee to overturn the House proposal.

The Black Hills Study

To further address charges that it was subsidizing the industry, the Forest Service prepared a detailed study of the Black Hills National Forest. Although the Forest Service considered this a "well-managed" forest, its cash flow study found that the Black Hills "had the greatest total negative cash flow" of any national forest.[29] According to the report, this negative cash flow was due to poor markets, problems with the Forest Service accounting system, the need to harvest timber to produce multiple-use benefits, and poor appraisal practices.

Poor markets had little to do with the negative cash flow. Prices had fallen considerably by 1983, the last year covered in the Forest Service's cash flow analysis. However, prices had never been higher than they were in 1979, which was also used in the cash flow study. The forest's cash flow was negative in both years.

The problems with the accounting system need closer attention. "Historically," said the report, "Congressional appropriations have been more generous to the timber function than to most other functions. The result has been that substantial portions of the cost of other resources, uses, and activities have been funded as though they were part of the timber program."[30] Black Hills managers estimated, for example, that "about 2 percent of their timber funding is used to support land management planning, and . . . another 2 to 4 percent of their timber funding is actually used in support of other resources programs."[31]

This total of no more than 6 percent appears insignificant. But the Forest Service also claimed that about 30 percent of the sale preparation costs were actually spent to support other resources. All land line location costs had been attributed to timber in 1983, and close to half of the road maintenance costs had been charged to timber. Together, these two items accounted for 82 percent of the support costs that the Forest Service claimed should really not be counted against timber. The remaining 18 percent included costs for such items as recreation, soil and water, and wildlife management.[32]

Claims that these costs should not be charged to timber are questionable. Land line location takes place primarily to prevent timber purchasers from trespassing onto adjacent private lands. Although the Forest Service maintains that land lines "will need to be determined at some point even if timber were never sold,"[33] in fact little effort is ever made to locate land lines outside of timber sales. The fact that this "should be done" does not mean it would ever be a high enough priority to be funded without timber sales.

Road maintenance is another cost that is clearly attributable to timber. Log truck traffic causes major damage to roads, and without the timber program road maintenance costs would be significantly less — as would the number of roads needed to manage the Forest.

Many of the other "support" costs do not actually benefit other resources but instead merely mitigate the damage timber can do to them. For example, a large share of the 18 percent is for cultural resource surveys to ensure that roads and timber cutting do not destroy historic and prehistoric sites.

In fact, virtually all of the "multiple-use benefits" identified by the Black Hills report were actually mitigation measures. For example, the study noted that sale preparation costs were increased by the need to protect visual quality in much of the forest.[34] However, the visual quality would be protected without the timber sales, so any costs imposed to maintain that quality should be charged strictly to timber.

The only serious criticism of Forest Service policies was contained in the discussion of appraisal practices. "In 1983," noted the study, "the Forest redesigned their appraisal process incorporating several recommendations" made by a joint Forest Service-industry committee. As a result, minimum bid prices fell by at least $32 per thousand board feet. Since much timber sold for the minimum bid price, the Forest Service essentially lost over $30 per thousand board feet after these changes were made.[35]

The Black Hills study team compared recent sales with the 1983 Black Hills Forest Plan, which claimed that timber would produce high benefits. The team found that the plan had overestimated timber prices by 100 percent and underestimated timber-related costs by 33 percent.[36] Despite these errors, the team accepted without question the plan's claims that timber management would produce high multiple-use benefits.

A Proposed Timber Sale Accounting System

In 1985, the Forest Service again faced criticism from the House Appropriations Committee. The agency told Congress it was making efforts to reduce costs and that it would increase income by increasing the "base rate" for timber — the rock-bottom minimum bid price — to ensure that sale preparation and administration costs, at least, would be covered. Since these were as high as $30 per thousand board feet on some forests that often sold timber for under $5 per thousand board feet, this represented a major change in policy. But Senator McClure successfully pressured the Forest Service to drop this proposal.

However, Congress appropriated $400,000 for the Forest Service to prepare a "timber sale accounting system" that would help determine exactly how much timber loses money.[37] The Forest Service had long claimed that the cash flow methods used by NRDC and the GAO were inappropriate. The House Appropriations Committee wanted to give the Forest Service a chance to determine the methods it thought would be correct.

The Forest Service presented its *Timber Sale Program Information Reporting System* to Congress in April 1987.[38] Rather than review individual timber sales, the system calls for comparing the budgets and collections for each national forest at the end of each year. The system also presents two separate accounting methods, one — a so-called "financial report" — reviewing the actual timber sale program, and the other — a so-called "economic report" — reviewing the net value of investments in second growth on the acres whose timber had been sold. The financial report is a cash flow analysis and will be discussed here, while the economic report is a bare land analysis and will be discussed in chapter 3.

The accounting system contains numerous flaws and unjustified assumptions, nearly all of which would make timber sales appear more worthwhile than they really are. The financial report ignores or "adjusts" major costs to make them appear negligible. For example, the Forest Service proposes to use a complicated formula to depreciate road and reforestation costs. The result is a serious underestimate of the actual costs attributable to timber sales. Test results from the Coconino National Forest (AZ), for example, count only $522,000 of road and reforestation costs against 1986 timber receipts.[39] Yet the Coconino typically spends about $1 million per year on reforestation and between $0.5 and $1 million per year on timber-related road construction.[40]

The depreciation formula is supposed to account for the fact that roads are an investment which will pay off in future, as well as current, timber sales. The formula used by the Forest Service, however, arbitrarily assumes that future timber sales will justify most road construction. No attempt is made to evaluate whether new roads will actually provide access to sufficient timber to pay their costs.

Reforestation costs probably should not be counted against timber in the financial report at all. Doing so implies that reforestation is an operating cost, not an investment in the future. By counting a fraction of reforestation costs here, the Forest Service avoids counting them at all in the economic report. As chapter 3 describes, the failure to count reforestation costs in the economic report results in a serious overestimate of the value of investments in future timber growth.

Land line location is another cost partially or completely disregarded by the financial report. Most land line location takes place exclusively to prevent timber purchasers from trespassing onto adjacent lands. While the Forest Service believes that land line location "is a basic cost of land stewardship,"[41] in fact it is such a low

Table 2.1 Cash Flow Analyses, Flathead National Forest, 1986

	Simple Cash Flow	*Draft TSPIRS*	Effects on Treasury"
	Receipts		
Timber	$2,318,870	$2,318,870	$2,318,870
Purchaser Credits	939,859	939,859	0
Knutson-Vandenberg Deposits	2,223,183	0	0
Total Receipts	$5,481,912	$3,258,729	$2,318,870
	Costs		
Sale Prep. & Admin.	$1,414,200	$1,414,200	$1,414,200
Timber Resource Planning	238,900	238,900	238,900
Silvicultural Examination	340,400	340,400	340,400
Appropriated Reforestation	387,300	348,570	0
Appropriated Timber Stand Impr.	298,000	298,000	0
Tree Improvement	229,000	229,000	0
K-V Expenditures	1,102,200	0	0
Appropriated Road Construction	2,469,200	1,492,998	2,469,200
Purchaser Credit Roads	939,859	0	0
Road Maintenance	0	0	307,600
Land Line Location	0	0	113,800
General Administration	0	882,307	882,307
Payments to Counties	0	0	1,370,478
Total Costs	$7,419,059	$5,244,375	$7,136,885
Net Receipts	-$1,937,147	-$1,985,646	-$4,818,015

The *Simple Cash Flow* column counts timber, road, and reforestation costs against timber. The *Draft TSPIRS* column approximates the results of using the Forest Service's draft *Timber Sale Program Information Reporting System.* The *Effects on Treasury* column compares income with costs to the U.S. Treasury.
Source: CHEC, "Review of the Flathead Forest Plan and Final EIS," (Eugene, OR: CHEC, 1987), p. 2.

priority that it would probably never be done were it not for the need to protect the Forest Service from timber trespass.

Despite these problems, evaluation of the financial report shows that many forests frequently lose money on timber. For example, the Flathead (MT) National Forest is rarely identified as a sales-below-cost forest. Yet, using available data to complete as much of the financial report as possible, the forest is estimated to have lost nearly $2 million on timber in 1986 (table 2.1).[42]

As proposed, the financial report of the timber sale accounting system will fail to identify many inefficiencies in national forest timber management. Money-losing sales on forests that tend to make money will be obscured because the system totals benefits and costs for entire national forests rather than for individual acres. The fact that the timber in most roadless areas is worth far less than the costs of roads needed to reach the timber will be obscured for the same reason.

Unfortunately, interminable debates about a cost accounting system overemphasize the importance of cash flow. While negative cash flow can indicate problems, by itself it provides little conclusive information. A forest with negative cash flow may still be efficiently managed. For example, many eastern forests were

Figure 2.2 Timber Cash Flow Between 1974 and 1978

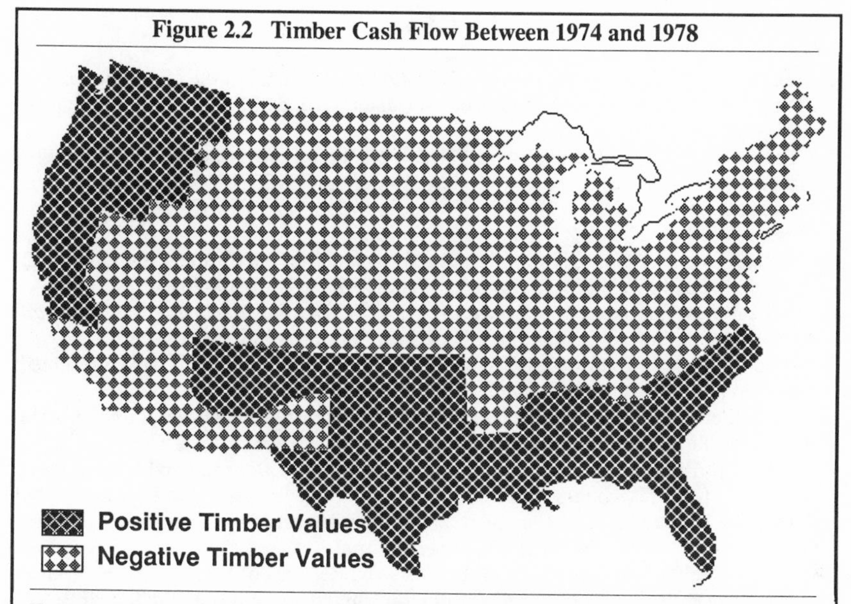

Forests with positive cash flow are almost exclusively located in the Pacific Northwest and the southern coastal plain. Since 1980, when timber prices fell, some of the forests shown in the positive regions on this map have returned negative cash flows.

Source: Thomas J. Barlow et al., *Giving Away the National Forests* (Washington, DC: Natural Resources Defense Council, 1980), appendix one.

largely denuded of merchantable timber before being purchased by the Forest Service. Such forests could not produce positive returns for many years, yet management investments may be worthwhile.

The fact that so many forests in the West, most of which contain large stocks of merchantable timber, have negative cash flows indicates that there are serious problems with Forest Service management in general. Further analysis is needed to address these problems, and cash flow can be a legitimate part of such an analysis.

A cash flow study preferably should take place on a per-acre basis, not per forest or per state. Acres should be grouped into units with similar costs and benefits. The costs of accessing each group and preparing timber sales would be compared with the expected return from such sales. As indicated by figure 2.2, such a per-acre cash flow analysis would find that millions of acres that lose money on timber are nevertheless being managed for timber by the Forest Service.

The fact that the net benefits for a particular group are negative does not necessarily mean that timber management is inappropriate. A prudent investor might be willing to lose money selling existing timber if the value of second growth were high enough to justify such sales. This possibility will be discussed in chapter 3. Alternatively, the benefits to nontimber resources might be great enough to justify losing money on timber sales. This possibility will be discussed in chapter 5.

Notes

1. Ken Gold and Dave Seifert, *Ten Successful Organizations and How They Do It* (Washington, DC: Productivity Resource Center, Office of Personnel Management, 1981), p. 1.
2. "The Role of 'Below Cost' Timber Sales in National Forest Management," unpublished paper on file at the Forest Service, Washington, DC, 16 August 1984, p. 7.
3. Ibid., p. 2
4. Ibid., p. 7.
5. Randal O'Toole, "Changing of the Guard — Interview With the New Chief and Associate Chief of the Forest Service," *Forest Watch* 7(10):15–20.
6. USDA Forest Service, *Timber Sale Program Information Reporting System Final Report to Congress* (Washington, DC: Forest Service, 1986), pp. 67–68.
7. David C. Iverson, "Below-Cost Sales: Much Ado About Something," paper presented to forest planners gathered in Ogden, UT, 26 November 1985.
8. "Forest Service Pricing Mechanism for National Forest Timber Sales," unpublished report on file at Natural Resources Defense Council, Washington, DC.
9. William F. Hyde, "Timber Economics in the Rockies: Efficiency and Management Options," *Land Economics* 57(4):630–637.
10. Thomas J. Barlow, et al., *Giving Away the National Forests* (Washington, DC: Natural Resources Defense Council, 1980), appendix one.
11. Gloria E. Helfand, "An Analysis of the Costs and Receipts of National Forest Timber Sales," master's thesis on file at Washington University, St. Louis, MO, p. 102.
12. "Role of 'Below Cost' Timber Sales," p. 5.
13. CHEC, *Economic Database for the Greater Yellowstone Forests* (Eugene, OR: CHEC, 1987), p. 10.
14. V. Alaric Sample, Jr., "Below-Cost Timber Sales on the National Forests," issue brief on file at the Wilderness Society, Washington, DC, July 1984, pp. 7–9.
15. Ibid., p. 12.
16. Robert E. Wolf, "Timber Sale Income and Expenditures, Tongass National Forest, Alaska," unpublished paper prepared for the Subcommittee on Mining, Forestry, and BPA of the House Interior Committee, Washington, DC, 8 August 1983, p. 1.
17. Robert E. Wolf, "State-by-State Estimates of Situations Where Timber Will Be Sold by the Forest Service at a Loss or a Profit," unpublished paper prepared for the Subcommittee on Interior Appropriations of the House Appropriations Committee, Washington, DC, 7 March 1984, p. 1.
18. Ibid., p. 11.
19. General Accounting Office, *Congress Needs Better Information on Forest Service's Below-Cost Timber Sales* (Washington, DC: GAO, 1984), p. 10.
20. "Role of 'Below Cost' Timber Sales," p. 1.
21. Ibid., pp. 2–3.
22. Ibid., p. 3.
23. Ibid., p. 4.
24. Senate Report #90-893, 94th Congress, second session, p. 2.
25. 36 *Code of Federal Regulations* 223.4(a).
26. *Forest Service Manual* 2421.3.
27. *Forest Service Manual* 2421.4.
28. "Analysis of Timber Sales in Response to HR-5973," unpublished paper on file at Forest Service office, Washington, DC, 23 July 1984, 13 pp.
29. "Analysis of Cash Flow in the Timber Sale Program on the Black Hills National Forest," unpublished paper on file with the USDA Forest Service, Washington, DC, p. 1.

30. Ibid., p. 8.
31. Ibid., pp. 5–7.
32. Ibid., p. 7.
33. Ibid., p. 8.
34. Ibid., p. 5.
35. Ibid., p. 16.
36. Ibid., pp. 20–21.
37. Public Law 98–473.
38. USDA Forest Service, *Timber Sale Program*.
39. "Draft Statement of Revenues and Expenses," memo on file at Coconino National Forest supervisor's office, Flagstaff, AZ, p. 90.
40. Budget statements for Region 3 forests for 1985 and 1986.
41. USDA Forest Service, *Timber Sale Program*, p. 67.
42. CHEC, "Review of the Flathead Forest Plan and Final EIS" (Eugene, OR: CHEC, 1987), p. 2.

Chapter Three
Poor Investments

Suppose that Lee Iacocca sold $100,000 worth of stock in Chrysler to buy, on the advice of his friend, the former chief of the Forest Service, stock in a large timber company. A year later, the price of the timber company stock has fallen from $50 to $30 per share, and Iacocca sells before it gets any worse. "Congratulations," says the chief. "You made a $60,000 profit."

"What profit?" says Iacocca. "I had $100,000, and now I have $60,000. I lost $40,000."

"No, you didn't," says the Chief. "The original $100,000 doesn't count because it was already paid for by the sale of your Chrysler stock."

"Actually, I lost more than $40,000," grumbles Iacocca. "The value of Chrysler stock grew by 10 percent in the last year. If I had left my money in Chrysler, I'd have $110,000 instead of $60,000."

"Oh, no!" urged the chief. "We never use a 10 percent interest rate when comparing our investments — trees just don't grow that fast. We use only a 4 percent interest rate. At 4 percent, you would have made $4,000 if you had left your money in Chrysler. Since you made $60,000 from selling the timber company stock, you made fifteen times as much as if you had not bought the timber company stock."

This imaginary chief's logic may be convoluted, yet it is repeated almost daily by forest and timber sale planners within the Forest Service. Their attitude toward reforestation and other costs reflects a long-held forester belief that timber management is cost-free. A bumper sticker posted in the Region 1 regional office proclaims, "Wood Is Good — Use It and Nature Renews It." Obviously, if nature does the renewing, people incur no costs.

Yet timber management is far from free. Although the Forest Service spends hundreds of millions of dollars each year to grow new stands of trees, it refuses to consider whether these investments are worthwhile. Instead, it says the investments were "paid for" by the returns from sales of existing timber, and thus there is no need to compare the costs of reforestation against the value of young growth timber. What little investment analysis takes place uses a 4 percent interest rate, which is significantly lower than the one used by the Forest Service's competitors in the forest products industry.

Bare land value studies have reasonably asked whether Forest Service investments are worthwhile. Some studies use higher interest, or *discount* rates, than 4 percent, but even when 4 percent is used, most studies find that national forest timber investments fare poorly.

Investments in Management

Forest Service timber investments include reforestation, thinnings, herbicides, fertilization, and other activities designed to make trees grow faster. The largest share is spent on reforestation. To ensure that new stands of trees are rapidly established to replace those recently harvested, many national forests raise seedlings in nurseries and plant them by hand. Without hand planting, harvested areas may still be naturally reforested by seed falling from adjacent timber. Hand planting only ensures that a somewhat higher percentage of the harvest units will be reforested within five years of the harvest.

Many parts of the arid West suffer from late-summer droughts in which exposed soils reach temperatures high enough to kill seedlings. To protect seedlings in clearcuts, the Forest Service often places *shade cards* — shakes of cedar to shade seedlings from the sun — over them. Other protective measures include wrapping seedlings in plastic mesh to protect them from deer and other animals.

Some forests have a program of selecting the best trees on the forest for seed and using that seed (or seed produced by progeny of the selected trees) to produce seedlings. The Forest Service claims that such a "genetic selection" program is likely to increase future growth by 10 percent or more.[1]

Herbicide applications are also made as an investment in future stand growth. Many clearcuts are sprayed with chemicals such as 2,4-D and Roundup immediately after harvesting and again a few years after seedlings have been planted. Such measures are designed to give conifers a competitive advantage over less desirable trees and shrubs. Without some form of brush control, conifers may take an additional ten to fifty or more years to fully stock a clearcut and be free to grow.

Although hand planting and herbicide spraying are used to ensure that harvested acres are stocked with enough trees to use the full productive capacity of the site, overplanting and natural seeding often lead to more than the optimum number of trees per acre. An overstocked stand might produce the same volume of timber as a less heavily stocked site, but the diameter of the trees will be smaller, making them less valuable. The Forest Service thus often *thins* such stands by cutting poor-quality or crowded trees. Thinnings between the ages of ten and thirty usually produce no usable wood, so they are called *precommercial thinnings*.

Older stands may be *commercially thinned*, meaning that the harvested trees are removed from the forest and used as poles, fence posts, pulp, or possibly even cut into lumber. However, the volumes produced by commercial thins are small and the value low, so many commercial thins lose money. The Forest Service justifies this by the fact that the remaining trees will grow larger in diameter and will be more valuable when they are finally harvested.

Fertilization is still only experimental in many national forests. Many soils are short on available nitrogen, and the Forest Service can apply nitrogen in the form of urea, a petroleum by-product, from helicopters or on the ground. Fertilization is expected to greatly stimulate the growth of timber growing on nitrogen-poor soils.

Another action designed to grow more timber is reforestation of old brush fields. The brush fields may have been caused by fires or may be clearcuts that were not successfully reforested. Heavy use of herbicides followed by hand planting of seedlings is often expected to add to future timber supplies.

Altogether, reforestation, precommercial thinning, herbicide release, fertilization, and the use of shade cards and other protection for seedlings may cost close to $1,000 per acre.[2] The costs are even higher when managers attempt to reforest brush fields, since the cost of removing existing vegetation is high.

A proper economic analysis would compare these costs with the present value of the returns produced by the activities. Cash flow studies, of course, do not do this. Some of these studies counted the money spent to reforest some acres of a forest, for example, against the value of timber sold from other acres of the forest. Yet the cutting of timber on one part of the forest is unrelated to the reforestation of another part of the forest.

Accounting for the Future

An important but often misunderstood tool for this type of analysis is the discount rate. This rate is used to compare future benefits and costs with those produced today. The discount rate recognizes that everyone places more value on the near future than on the distant future. For example, when asked whether they would prefer to have $1 today or $10 in ten years, most people will take the dollar today. Since $1 invested today will not grow to be $10 unless the interest rate paid on the investment is more than 26 percent, this indicates that most people have a personal discount rate of at least 26 percent.

There are several reasons for preferring goods now. Young people may expect to have rising incomes in the future and thus believe that $10 in ten years will just not be as important to them as $1 is today. Retired people on fixed incomes know that there is a high possibility that they will no longer be alive in ten years and so also prefer the money today. There is also an element of uncertainty: The payer may not be alive or may go bankrupt in ten years, making it impossible to ever collect the $10 in the future.

The discount rates of investors tend to be well under 26 percent. Investors typically use one of two methods to determine their discount rate. Investors who are not using their own money will need to borrow to make the investment. Money is loaned at a specified rate of interest, and if the investment does not return at least that rate, the investor will lose money. Thus, the *cost of borrowing money* becomes the minimum discount rate used.

Other investors know that they have the option of certain investments that will guarantee them a particular rate of return. For example, money market funds have recently been paying 6 percent. Investments that do not pay at least 6 percent are thus considered unfavorably, and the *alternate rate of return*—6 percent in this case — becomes the discount rate.

Planners sometimes calculate the present value of future benefits and costs using several different discount rates. This does not mean that a discount rate which produces a predetermined answer can be selected. *The discount rate is the cost of money*, and it can't be changed any more than the price of groceries, a car, or a house can be changed.

Discount rates are similar to compound interest, but in reverse. One dollar invested at 5 percent compound interest will grow to $1.63 in ten years. Invested at 10 percent it will grow to $2.59 in ten years. This can be calculated using the formula

$$FV = PV \times (1 + i)^t$$

where FV = future value
PV = present value
i = the interest rate expressed as a decimal
t = the number of years of the investment

Discounting reverses this formula, so that a value of $2.59 in ten years is worth $1 today at 10 percent interest. This amount is calculated with the formula

$$PV = \frac{FV}{(1 + i)^t}$$

When considering benefits and costs, economists prefer to use *constant dollars*, or dollars adjusted for inflation. Since inflation affects the price of all goods equally, future inflation is simply assumed to be zero. Inflation has not been zero in the recent past, so when historic revenues or expenses are counted, these values are adjusted using the consumer price index or another index of inflation.

Although inflation affects all prices equally, some goods increase in value faster than others as the demand for those goods increases relative to the supply. In some cases, changes in taste may increase the demand for something, such as cross-country skiing. In other cases, reductions in supply may make an item more scarce and thus more valuable. The rate of change in the value of a good or service can be expressed as a percent, like a discount rate, and is called a *trend.*

Bare Land Values

Studies of bare land value — sometimes known as *soil expectation value* or *soil rent* — of forestland apply these discounting principles to second-growth timber management (table 3.1). Lands are assumed to be bare because of a fire, clearcut, or other cause. Reforestation and other investment costs are discounted where appropriate and totaled. The value of timber that will be grown and harvested in the future is also discounted and totaled.

One of the first widely publicized studies of the bare land value of a national forest was made in the University of Montana professors' report on clearcutting in the Bitterroot Forest.[3] The report assumed that reforestation cost $50 per acre, which it said was "conservative." No other costs were assumed. After 120 years, 20,000 board feet were assumed to be available for harvest, a volume that was considered "optimistic for most sites on the Bitterroot."

At a 5 percent discount rate, which the report called "conservative," "the stand at harvest would have to be worth $17,445 per acre," or "$872 per thousand board feet." If timber values were only $25 per thousand board feet, which was typical for

Table 3.1 Sample Bare Land Value Analysis

	Year	Cost/ Acre	Revenue /MCF	MCF /Acre	PV/ Acre
Reforestation	1	$285			-$274
Release	5	98			-81
Precommercial Thin	15	178			-99
First Commercial Thin	50		$140	1	16
Second Commercial Thin	80		356	1	20
Final Harvest	110		663	5	46
Total		$561		4,015	-$372

Comparison of discounted costs with discounted revenues for a mixed conifer site on the Bitterroot National Forest. Future values are discounted at 4 percent per year to obtain the present values per acre (PV/Acre) shown in the last column.
Source: CHEC, "Review of the Draft Bitterroot Forest Plan and EIS" (Eugene, OR: CHEC, 1985), p. 2.

1970, "the yield would have to be 697,000 board feet per acre," a volume which the report said was "obviously impossible."

The report concluded that clearcutting and planting "cannot be justified as an investment for producing timber on the Bitterroot National Forest." The writers recommended that the use of clearcutting and planting "should bear some relationship to the capability of the site to return the cost invested."[4] Although the Forest Service has stopped terracing slopes after harvesting, it still practices clearcutting and planting in the Bitterroot Forest.

William Hyde used similar techniques but with more sophisticated data in his study of the San Juan National Forest. Hyde assumed that timber prices will increase at 2.5 percent per year, and he discounted future benefits and costs at 7 percent and, alternatively, at 10 percent. As costs, Hyde counted sale administration, general administration, and fire protection. In addition, he noted that most lands were regenerated naturally but that hand planting costing $367 per acre was needed slightly more than 10 percent of the time. With this information, he calculated the price that timber must be worth today to justify investing in second-growth using either natural or artificial reforestation.[5]

At the 7 percent discount rate, Hyde calculated that prices today must be over $110 per thousand board feet to justify natural reforestation and over $290 per thousand board feet to justify hand planting. At the 10 percent rate, timber today must be worth over $530 per thousand board feet for natural and over $1,000 per thousand board feet for artificial regeneration. Since most recent timber sales sold for $25 per thousand, Hyde concluded that "market criteria fail to justify continuing timber management."

The Forest Service has never used this type of analysis in any of its planning or budgetary processes. For example, the economic report of the Forest Service's proposed timber sale accounting system ignores the cost of reforestation.[6] Instead, this cost is charged in the financial report against the value of trees that were just harvested. At first glance, this might seem to make sense. But in reality the fact that trees are cut from an acre does not mean that the Forest Service has to spend huge amounts of money reforesting that acre. Reforestation is done to produce timber in the future, not the timber that was cut last year. Therefore, its cost should be compared with the anticipated return from the timber into which the planted trees will grow.

This is especially important when alternate means of reforestation are considered. Suppose that natural reforestation, the costs of which are negligible, is compared with hand planting, which can cost more than $500 per acre. On a good site, harvests of existing timber may produce revenues of $5,000 per acre, which would easily cover either hand planting or natural reforestation. But the reforestation method selected makes no difference to the revenue produced from timber harvests. Instead, it is the returns from the young growth stand that vary with a change in reforestation methods, and it is these returns that must be compared with costs when deciding whether to use artificial or natural reforestation.

Yet the Forest Service rarely compares returns with the costs of investments. For example, forest planning rules require planners to estimate the present net value of various regimes, presumably including regimes with and without thinnings, herbicides, and other practices. Planners were to identify the management techniques that would produce the highest net benefits.[7] The results of this analysis played no part in any decision or any further planning step, however, so most planners simply did not bother.

More recently, the Forest Service directed managers to estimate the return on investments in reforestation, thinnings, and other intensive management practices in order to form budgeting priorities, but managers are specifically directed not to use discount rates. The result is greatly exaggerated returns. Managers are also to assume that failure to reforest areas proposed for hand planting will result in "*no* volume" (emphasis in original) produced by the site — ignoring the possibility that

Table 3.2 Bare Land Values of Selected Oregon Forest Types

National Forest	*Forest Type*	*5%*	*6-7/8%*	*10%*
Mt. Hood	Douglas-Fir Site 3	$1,565	$447	-$17
Mt. Hood	Douglas-Fir Site 4	588	55	-108
Mt. Hood	Pacific Silver Fir	320	-58	-131
Siskiyou	Douglas-Fir	622	-30	-253
Deschutes	Ponderosa Pine	549	26	-104
Deschutes	Lodgepole Pine	71	-68	-114
Deschutes	Mixed Conifer	694	-79	-92
Wallowa-Whitman	Ponderosa Pine	-36	-147	-168
Wallowa-Whitman	Lodgepole Pine	-83	-152	-170
Wallowa-Whitman	Mixed Conifer	-21	-150	-172

Values shown are in dollars per acre discounted at three different discount rates. Timber prices have fallen and costs have increased since these calculations were made, so the bare land values shown are probably higher than realistic.

Source: Randal O'Toole, *A New Reality: Timber Land Suitability in Oregon National Forests* (Eugene, OR: CHEC, 1979), p. 12.

natural reforestation might be nearly as successful as hand planting.[8]

While the Forest Service has never correctly evaluated the efficiency of timber investments, calculations show that investments on most national forests will produce a low rate of return that would be unacceptable to private landowners. In 1979, CHEC reviewed inventory data for all thirteen Oregon national forests. These data estimated the number of acres of each forest by forest type and productivity class. Interviews with timber managers on each forest produced estimates of the value of the timber and the costs of reforestation, precommercial thinning, and other planned management activities, as well as Forest Service estimates of the yield of each forest type and productivity class. This information was used to perform a bare land analysis for the major forest types in all Oregon forests (table 3.2).

CHEC concluded that most national forestlands in eastern Oregon were not capable of producing a 5 percent rate of return on investments in reforestation.[9] Most acres in northwest Oregon forests could produce this return, but many in southwest Oregon could not. Higher discount rates made the results even more dismal: At 6-7/8th percent, the rate recommended at that time for federal water projects by the Federal Water Resources Council, most lands lost money. Only a few acres of the most productive land could produce 10 percent, the rate recommended by the Office of Management and Budget (OMB).

Which is the appropriate rate? As CHEC completed its study, the Forest Service received reluctant approval from OMB to use a 4 percent rate. Since this rate is lower than any considered by CHEC, national forest officials in Oregon ignored the results of CHEC's report.

But 4 percent is far too low. Despite claims by timber industry officials that private lands are "better" managed than the national forests, the Forest Service clearly invests in less productive land than the industry. For example, long-standing Forest Service policies call for managing any timberland capable of growing 20 cubic feet of wood per acre per year.[10] In 1982, this policy was relaxed so that forests could manage timber on land that grew as little as 12 cubic feet of wood per acre per year.[11]

In contrast, timber industry officials say that 80 cubic feet is generally the lowest productivity they would find acceptable.[12] Although Oregon's Forest Practice Rules require reforestation after clearcutting, the rules do not require any reforestation on lands producing under 50 cubic feet per acre per year, indicating that Oregon's Board of Forestry does not believe that less productive land is economically worth managing.[13]

The Forest Service is overinvesting in other areas as well. To reforest brush fields, it sprays herbicides and burns brush in preparation for planting with conifers. This practice, which is applied to low-productivity forestlands in the hot, dry valleys of southwest Oregon, costs $600 to $1,000 per acre.[14] Yet the timber industry applies the same techniques only to the most productive forestlands in Oregon's Coast Range — the nearest thing to a rain forest found in the state. Such overinvestments imply that the Forest Service's discount rate is too low.

An extensive body of economic literature discusses the appropriate discount rate to use for public projects. Richard Mikesell, an economist at the University of

Oregon, reviewed the literature in 1980 to see if there is any consensus. He concluded that "the rate of discount for evaluating public projects should be based on the opportunity cost of capital in the private sector."[15] In other words, when a public agency is competing directly with private operators, the public agency should use the same rate as "the before-tax rate of return on projects in the same risk class" of the private sector. The only possible adjustment would be a slight reduction to account for the fact that the federal government, which is much larger than any private company, has less risk to fear than a private investor. In short, Forest Service timber planners should use the same rate as, or perhaps a slightly lower rate than, the pretax rate used by the timber industry.

According to a 1985 survey of over forty major timber companies, the industry used real pretax discount rates averaging about 7 percent between 1976 and 1980. With the sudden increase in interest rates at the end of 1980, the average discount rate for these companies increased to over 12 percent (figure 3.1).[16] By 1984, the rate had fallen to 11 percent. Of course, at these higher rates the industry stopped making many of the investments it had been making in the 1970s. But most companies remained in business and continued to reforest their most productive lands. Interest rates may come down, but it is unlikely that timber company discount

Figure 3.1 Timber Industry and Forest Service Discount Rates

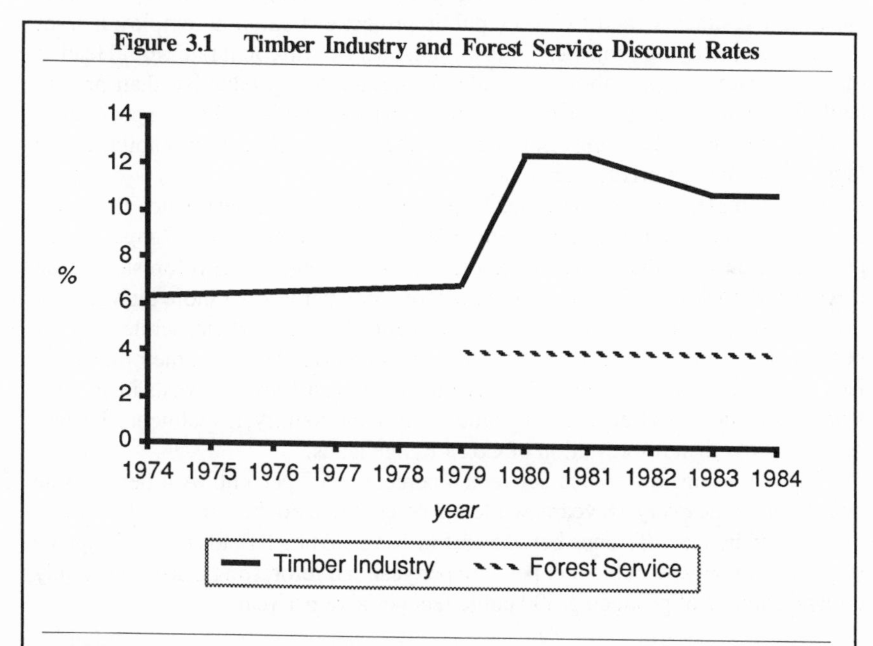

Rates shown are real rates before taxes. The Forest Service's lower rate implies that it will make investments in national forest timber management that produce poorer returns than investments which are not being made, for lack of funds, by industry on private lands.

Source: Clair H. Redmond and Frederick W. Cubbage, *Capital Budgeting in the Forest Products Industry: A Survey and Analysis* (Athens, GA: University of Georgia, 1985), p. 17.

rates will ever fall below the levels used in the late 1970s, which was the most optimistic period in the history of the industry.

The Forest Service used a completely different method to arrive at its 4 percent rate. It asked the Federal Reserve Board what the *actual* rate of return had been from private investments in recent years. The Board estimated it was about 2.5 percent. Since this was an after-tax return, the Forest Service increased it to 4 percent to make it equivalent to a pretax return.[17] This method is inappropriate because no one intentionally makes a bad investment — yet many people do lose money on investments. Using the average rate of return instead of the desired rate of return results in an inappropriately lower discount rate than the one used by private industry.

The Forest Service's methodology was reviewed by only a few people, all of whom were within the agency itself. One of the economists who helped determine the rate said that a higher rate was not appropriate because "trees just don't grow that fast."[18] Yet most private companies were using much higher rates and still managed to find worthwhile investments in their lands — partly because their lands were, on average, more productive than the national forests.

Mikesell argued that the Forest Service should use the same rate as industry "to prevent the public sector from transferring capital from higher-yielding to lower-yielding investments" and to "force public project evaluators to employ market standards in justifying projects." A low Forest Service discount rate would lead to timber investments on national forestlands that are less productive than private lands that go unmanaged for lack of capital. To prevent this, Mikesell specifically recommended that the Forest Service use "the appropriate cost of capital in the logging industry on private forest lands."[19]

For example, the Oregon State Department of Forestry estimates that over 550,000 acres of potentially prime timberlands in the Oregon Coast Range are unproductive because they were not reforested after being cut. If reforested, these lands could produce 525 million board feet of wood per year — more than each of four Forest Service regions.[20] Over 75 percent of this is private, while only 11 percent is national forestland. These lands are unmanaged because their managers lack the funds to reforest them. Meanwhile, the Forest Service invests in conversion of low-site brush fields in many other parts of the country, investments that will return far less than reforestation of Coast Range lands.

There is a huge difference between 4 percent and 7 percent. A 4 percent rate doubles in value every 18 years, while a 7 percent rate doubles in just 10. This is represented by the difference between reforesting old brush fields on land capable of producing only 50 cubic feet per acre per year and reforesting brush fields only on land capable of producing 200 cubic feet per acre per year.

Harvest and Reforestation Decisions

Critics of CHEC's proposal to consider bare land values in planning said CHEC failed to account for the value of old-growth forests. Of course, the purpose of bare land valuation is to estimate the value of the land after the old-growth is gone. But economists at Oregon State University (OSU) and in the Forest Service believed that land was economically suitable for timber management if the value of the old-growth exceeded the losses from second-growth management. "For national forest lands, harvesting and reforestation are not separate issues," said a group of OSU professors, because the law requires reforestation after harvest.[21]

But the law itself may be economically questionable. A particular acre of timberland can fall into any one of four categories: (1) The existing timber can be profitably harvested,and investments in second growth will produce a reasonable rate of return; (2) both sales of existing timber and management of second growth will lose money; (3) existing sales lose money but second growth is a worthwhile investment; (4) the existing timber is valuable but second growth is a poor investment.

This last category is the one under debate. Most national forests in Oregon and Washington make money on sales of old-growth timber. Yet CHEC's study of Oregon forests indicated that some forestlands in the Northwest were incapable of producing a reasonable rate of return on second-growth management. In particular, the high-elevation forests of the Cascade mountains and the mixed conifer, lodgepole pine, and low-site ponderosa pine forests east of the Cascades are marginal or submarginal investments.[22] The Forest Service, supported by OSU agricultural economists, believes that those money-losing investments should be made simply because they were "paid for" by the returns from harvesting existing

Table 3.3 Bare Land Values on Selected Forests and Types

National Forest	*Forest Type*	*Management*	*PNV*
Beaverhead (MT)	Mixed Conifer	Low	-$48
Bitterroot (MT)	Douglas-Fir	Medium	-259
Chattahoochee-Oconee (GA)	Cove Hardwoods	Medium	-46
Chattahoochee-Oconee (GA)	White Pine	Medium	-24
Clearwater (ID)	Mixed Conifer	Medium	50
Clearwater (ID)	Mixed Conifer	High	-84
Flathead (MT)	Mixed Conifer	Medium	-46
Klamath (CA)	Ponderosa Pine	High	-27
Klamath (CA)	Ponderosa Pine	Medium	9
Nantahala-Pisgah (NC)	Cove Hardwoods	Low	-286
Nantahala-Pisgah (NC)	White Pine	Low	-75
Okanogan (WA)	Mixed Conifer	Low	-6
Okanogan (WA)	Mixed Conifer	Medium	-15
Okanogan (WA)	Mixed Conifer	High	-37
Plumas (CA)	Mixed Conifer	Low	-346
Plumas (CA)	Ponderosa Pine	Low	-522
Shasta-Trinity (CA)	Douglas-Fir	High	252
Shasta-Trinity (CA)	Mixed Conifer	Low	-138

Receipts and costs are discounted at 4 percent per year to calculate present net values. The few sites with positive bare land values represent the most productive sites of a few national forests.

Source: CHEC reviews of plans for indicated forests

timber. This is the same as the chief telling Lee Iacocca that he didn't lose money on his timber stock because the stock was paid for by the sale of Chrysler stock.

Reforestation is usually the major investment cost in growing a new stand of trees. Unless the land is very productive, as most private land is, the returns on reforestation investments are very low — usually even less than 4 percent on the national forests. By pretending that reforestation costs nothing because it is "paid for" by the receipts from existing timber harvests, the rate of return suddenly increases.

Forests in northern California, northern Idaho, northwest Montana, and the southern coastal plain also routinely collect more than they spend. But except for the coastal plain forests, CHEC's analyses of forest plans have found that all but the most productive lands in these forests are poor second-growth investments. Rates of return even for lands producing high-value species like ponderosa pine are typically 2 to 3 percent per year.

Table 3.3 shows the bare land value of a number of forest sites as calculated by CHEC using data from forest plans. Based on these and other calculations, figure 3.2 shows the returns that can be expected from national forests in various parts of the country. Comparing this figure with figure 2.2 shows that most national forests, including those in Alaska and throughout most of the Rocky Mountains, southern California, the midwest, and the Appalachian and other mountain regions of the East, both lose money on most of the timber they sell and produce less than a 4 percent rate of return on investments.

For example, the Forest Service argued that mismanagement prior to its ownership of forests in western North Carolina had left those forests in poor condition. Proper management, according to this line of reasoning, would result in highly productive forests that would make money. Yet a bare land analysis of these forests, using Forest Service timber values and yield tables, found no acres that could produce a 4 percent rate of return on reforestation investments. When the cost of cutting existing timber was added in, timber management was clearly inferior to the option of simply leaving the forests alone.

Figures 3.2 and 2.2 also show that some forests that make money on most sales of old-growth will lose money on investments in second growth. This includes many forests in eastern Oregon and Washington, northern California, and Idaho. Only the most productive sites in western Washington, northwestern Oregon, northern California, northern Idaho, and the southern Piedmont could claim to make money today and be attractive sites for investments in timber management. Very few forestlands fall into the category that loses money today but makes money in the future.

Negative bare land values can be made positive in three ways. First, reforestation and other costs can be reduced. This may require that reforestation take longer than five years or it may require a switch to cutting methods that are more likely to naturally reforest stands. Second, sites can simply be dismissed from the timber base after harvest. If the site happens to reforest itself naturally, it can be harvested later. If not, the return from harvesting existing timber is probably more productively invested in activities other than reforestation. Finally, lands with

Figure 3.2 Bare Land Values on the National Forests

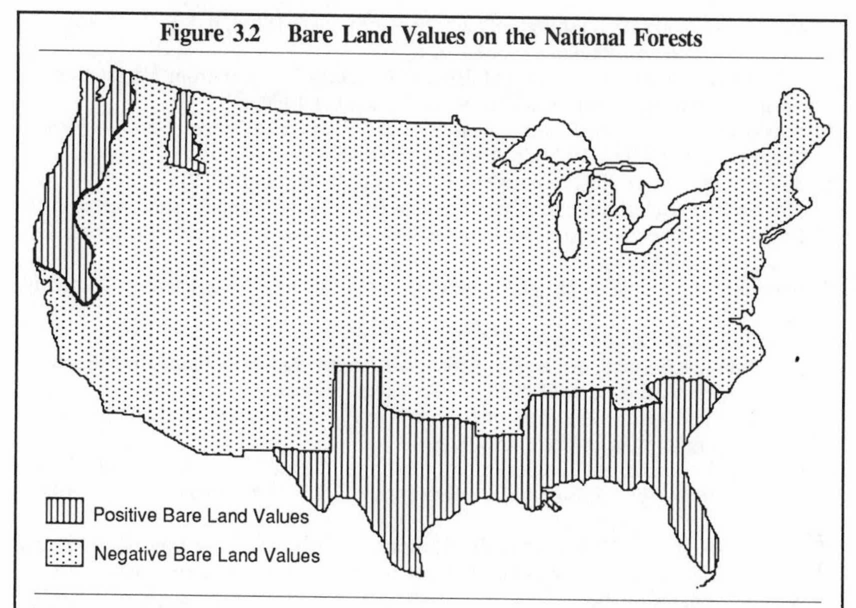

Most investments in second-growth timber on the national forests will produce negative returns when future values are discounted at four percent. Many specific forest types in the area indicated *positive* will actually have negative returns, while few forest lands in the area indicated *negative* will produce more than a 4 percent discount rate.

negative bare land values can simply not be harvested. This is certainly the best option where sales of existing timber lose money. It is also a worthwhile option to consider when existing timber makes money, but resource values that conflict with timber sales are also high.

Bare land analyses are even more unfavorable to Forest Service timber management plans than cash flow analysis. Negative cash flows are only a symptom of the problem — the real problem is actually much worse: The Forest Service is spending hundreds of millions of taxpayers' dollars on investments that would send any private landowner into bankruptcy.

Notes

1. USDA Forest Service, *Siuslaw National Forest Plan Draft Environmental Impact Statement* (Corvallis, OR: Forest Service, 1986), p. B-30.
2. USDA Forest Service, *Rogue River National Timber Resource Final Draft Environmental Impact Statement* (Corvallis, OR: Forest Service, 1978), p. J-2.
3. Arnold W. Bolle et al., *A University View of the Forest Service* (Washington, DC: U.S. Senate Committee on Interior and Insular Affairs, 1970), p. 21.
4. Ibid., p. 13.
5. William F. Hyde, "Timber Economics in the Rockies: Efficiency and Management Options," *Land Economics* 57(4):630–637.
6. USDA Forest Service, *Timber Sale Program Information Reporting System Draft*

Report to Congress (Washington, DC: Forest Service, 1986), p. 80.
7. 36 *Code of Federal Regulations* 214.14(b).
8. "FY 1989 Reforestation and TSI Budget Priorities," memo from Forest Service Washington office to regional offices, 11 September 1986, p 7.
9. Randal O'Toole, *A New Reality: Timber Land Suitability in Oregon National Forests* (Eugene, OR: CHEC, 1979), p. 12.
10. 36 *Code of Federal Regulations* 219.12(b)(1)(ii) (version of 1979).
11. "Capability: Determination of Commercial Forest Land," unpublished memo from Region 6 to Ochoco Forest Supervisor on file at Forest Service office in Prineville, OR, 2 December 1981, 1 p.
12. Interview with Charles Bingham, senior vice-president, Weyerhaeuser Company, May 1974.
13. *Oregon Forest Practice Rules* §24-501.
14. USDA Forest Service, *Rogue River National Timber Resource Final Draft Environmental Impact Statement* (Corvallis, OR: Forest Service, 1978), p. J-2.
15. Richard F. Mikesell, *The Rate of Discount in Public Decisions* (Washington, DC: American Enterprise Institute, 1980), p. 40.
16. Clair H. Redmond and Frederick W. Cubbage, *Capital Budgeting in the Forest Products Industry: A Survey and Analysis* (Athens, GA: University of Georgia, 1985), p. 17.
17. Clark Row, H. Fred Kaiser, and John Sessions, "Discount Rate for Long-Term Forest Service Investments," unpublished paper on file with the USDA Forest Service, Washington, DC, p. 6.
18. Clark Row, "Forest Service Budget Maximization: A Dissent," *Renewable Resources Journal*, 2(4):5–7.
19. Mikesell, *Rate of Discount*, p. 42.
20. Oregon State Department of Forestry, *Forestry Program for Oregon Supplement Number 2* (Salem, OR: Oregon State Department of Forestry, 1977), p. iii.
21. R. A. Oliveira et al., "Evaluation of 'A New Reality,'" unpublished memo on file at Forest Service office in Baker, OR, October, 1979, p. 1.
22. O'Toole, *A New Reality*, p. 15.

Chapter Four
Planning Follies

At the meeting between Chrysler's board of directors and the former chief of the Forest Service, one of the board members asks why the chief proposed changing Chrysler's system of pricing automobiles.

"Americans purchased 10 million cars last year," the chief answers, "but our studies show that the amount people will pay for cars will increase by 5 percent per year for the next ten years. Other auto manufacturers don't seem to be gearing up for this increase in demand, so Chrysler should be able to take advantage of it. The new pricing system is the best way to do it because it gives Chrysler the best chance to sell cars."

"Have you estimated the number of cars Chrysler can sell using your system?" asks Iacocca.

"Yes," says the chief, "those estimates are all found in my plan for the *renewable production of automobiles*, also known as the RPA plan."

"What if some dealers don't want to sell so many cars at such low prices?" asks a board member.

"Under the RPA plan, each dealer is given a quota of cars to sell each year. The quota steadily increases so that, by 1990, Chrysler has surpassed even General Motors in annual sales. The dealers will be required to file a *dealer plan* that shows how they will sell that many cars."

"Wait a minute," interrupts one of the board members. "What if prices don't increase at the rate you project?"

"The RPA plan and dealer plans will be reassessed in 1995. In the meantime, Chrysler shouldn't let temporary downturns in prices influence the plan."

"What happens if prices do increase but our costs increase at the same time?"

"That's one of the beauties of this system," responds the chief. "Since there is no relationship between the cost to Chrysler and the price of the car, Chrysler will be able to capture most of the market."

As suggested by this imaginary chief, the RPA and forest planning process is characterized by the lack of a realistic relationship between costs and returns. Instead, the RPA programs have forecast rapidly increasing demands for timber that generally have assumed that prices and costs remain constant. To meet this supposed demand, each national forest is assigned a timber target that increases over time.

The national forests attempt to meet these targets by allocating, in their forest plans, enough forestland to timber production to sell their share. Because timber is sold at a price agreeable to purchasers and with little regard to the cost to the Forest Service, the prophecy of increasing demand becomes self-fulfilling: Most of the timber offered for sale is sold.

As a result, millions of acres of money-losing timberlands are managed for timber, often without explicit public recognition in the forest plans. Instead, flaws in the overall planning process combine with specific flaws in individual plans to deceptively indicate that money-losing timber sales rarely, if ever, take place in the forests.

The most important procedural flaws are the low 4 percent discount rate, the failure to separately evaluate the economic value of existing stands from the bare land value, and the use of cost-efficiency rather than efficiency as the criteria for determining suitable timberlands. The 4 percent discount rate was discussed in chapter 3; the other two problems are discussed in this chapter.

In addition to these problems, many forest plans have specific flaws that introduce biases toward timber even when timber sales lose money. Most plans use timber prices that are much higher than purchasers are bidding or are expected to bid in the next several years. Many plans also use high price trends — assumptions that prices will rapidly increase over the next fifty years. The costs of roads and certain other activities are sometimes ignored or underestimated. Finally, many plans make unrealistic assumptions about the benefits timber sales provide to other resources. This last point is discussed in chapter 5, while the others are discussed in this chapter.

Economically Suitable Lands That Lose Money

A central issue in the determination of which lands are suitable for timber management is the Forest Service's decision to combine existing and second-growth timber values in its economic analyses. As Hyde's analysis of the San Juan National Forest revealed, the decision to manage land for timber is really two separate decisions: (1) the decision to harvest the timber now standing on the land and (2) the decision to grow a new stand of trees.[1] This distinction is not recognized in forest plans.

As pointed out in chapter 3, a piece of land can fall into any of four categories. Existing timber harvests and second-growth investments can both be profitable, they can both be unprofitable, or one can be profitable and the other unprofitable. This classification is never made by forest planners, because they do not calculate the value of existing timber separately from the return on second-growth investments. Instead, reforestation and other management investments are assumed to always follow timber sales. Calculations of economic value are made almost exclusively by a "FORest PLANning" computer program called FORPLAN.

To use FORPLAN, planners divide the forest into classes, called *analysis areas*, of similar benefits and costs. The benefits and costs of managing an acre of each analysis area are entered into the computer. The computer is then asked to propose the plan that produces the highest economic value, subject to limits such as a minimum level of timber sales.[2]

FORPLAN calculates, but does not normally print out, the value of sales of existing timber and the return on investments in second growth. Thus, planners rarely are aware of the two separate values. In most cases where existing sales pay off but second-growth investments do not, the return from existing sales is larger than the loss from second growth investments. Thus, when directed to maximize timber values, FORPLAN will generally propose to sell timber even though, after the existing timber is gone, continued management requires perpetual money-losing investments. In many cases, the sensible decision would be to mine existing timber and make no investments in second growth. The Forest Service is so strongly committed to sustained yield, however, that it refuses to consider such an option.

In other cases, the loss to other resources, such as recreation and wildlife, caused by timber sales is greater than the value of the existing timber. This loss could be partially mitigated by investing in second-growth management. But the option with the highest economic value may be no cutting at all. This option is rarely considered. For example, the FORPLAN model used for the final Shoshone Plan gave FORPLAN the choice of timber with high levels of recreation or no timber with very low levels of recreation. The highest-valued option — high levels of recreation with no timber — was excluded from the FORPLAN model.[3]

Thus, forest plans fail to consider two important and potentially efficient options: timber harvests with no follow-up reforestation, and no timber harvests with high outputs of other resources. The usual result is that FORPLAN proposes to harvest existing timber even though such harvests require perpetual money-losing investments in the future.

Future plans may recognize that those investments are not worthwhile. Such plans may propose to reduce timber sales to avoid the money-losing investments. Thus, even though today's plans are designed to ensure that nondeclining yields of timber are physically possible, actual yields will probably decline in the future for economic reasons. This phenomenon has been called the "declining even flow effect."[4]

Together, the low 4 percent discount rate and the failure to consider possible losses from second-growth management may lead some forests to propose inefficient investments in future timber management. On most forests, however, an even more serious problem is the use of timber targets rather than economic efficiency to determine the suitable timber base. Lands needed to meet the targets are considered "economically suitable" even if timber sales on those lands lose money.

Timber Targets

The source of the timber targets is a national plan called the *RPA program.* Under the 1974 Forest and Rangelands Renewable Resources Planning Act (RPA), the Forest Service prepares an RPA program every five years, describing the appropriations the Forest Service hopes to receive for the next five years and the forest outputs that would be produced with those budgets. RPA was expanded beyond use as a mere budgeting tool with the passage of the National Forest Management Act (NFMA) in 1976. Most of NFMA was an amendment to RPA, particularly the sections dealing with forest planning, but the law made no clear tie between the RPA plan and forest planning.

Such a tie was made by the Forest Service's planning rules adopted in 1979. The RPA program would define regional objectives for timber, grazing, and other resources. These objectives would be passed to each forest by regional plans or guides. At least one alternative in the forest plans would attempt to meet these objectives.

Although NFMA directed that physically and economically unsuitable forestland was to be withdrawn from the timber base before the annual sale levels were calculated by forest plans, the 1980 RPA objectives were developed before planners were able to assess either the economic value of the timber or the physical problems with timber management. Instead, forest supervisors were simply asked how much timber they estimated their forests could produce in the next ten years. RPA planners then assumed that improvements in technology would allow future access to lands that were, up to that time, considered unharvestable, and that future intensive management practices would also increase growth. Thus, the best estimates of local forest officials were often increased by 20 to 30 percent.

The possibly unrealistic nature of the timber objectives is compounded by the unbalanced nature of the commodity and amenity objectives. Timber and grazing objectives are "hard" — they are stated in terms of board feet and animal unit months, quotas that are easily measurable and directly under the control of forest managers. Wildlife, recreation, and watershed objectives are much "softer." Some, such as acre-feet of water and recreation visitor days, are mostly beyond the control of forest managers. Others, such as acres of wildlife habitat improvement, may not have any relationship to actual wildlife populations.

The Forest Service insists that RPA objectives are not firm targets forest planners must meet. Instead, forest supervisors are free to "negotiate" the objectives with regional foresters if they believe that some objectives are too high. If the timber objective for a particular national forest is reduced through negotiations, the regional forester has a choice between increasing the objective for another forest or going to the chief and negotiating a new regional target.

Yet forest personnel have always been rated for their success in meeting objectives. It is not surprising that many planners speak of pressure to meet timber targets and that many forest plans do, in fact, meet these targets. For example, the draft Klamath Forest Plan, published in 1983, called for a departure from nondeclining flow. This unprecedented step was needed, according to a memo by the forest supervisor, to meet the RPA target.[5] Since NFMA does not authorize the Forest Service to depart from nondeclining flow simply to meet an RPA target, the memo listed other reasons that would be used to publicly justify the departure — reasons developed after the decision to depart had been reached. Although most other forest plans also meet their RPA timber targets, officials often claim that the plans are designed to meet other criteria and that meeting the targets is merely a coincidence.

The RPA timber objectives are based almost entirely on Forest Service esti-

Figure 4.1 Demand Curves

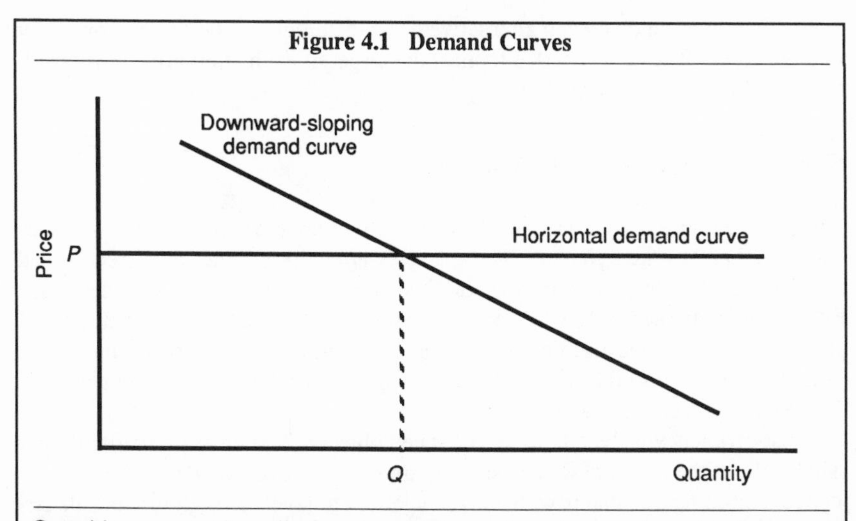

Quantities consumed usually decrease with increasing prices, so demand is downward sloping. But Forest Service planners assume that prices will stay constant no matter how much timber a national forest produces. This is called a horizontal demand curve. At the same time, they often say that "the demand for timber is *Q*" where *Q* is some fixed number. This seems to assume that demand will be *Q* no matter what the price is. In reality, *Q* will change with changing prices.

mates of the ability of the national forests to supply timber, but many planners argued that the objectives somehow represented the demand for timber. To planners, the need to meet this demand justified including thousands of acres of money-losing land in the timber base.[6]

In reality, demand is a curve representing the quantity of a good that people will buy at various price levels (figure 4.1). A single point on that curve, such as an RPA timber objective, provides very little information, particularly if it is not accompanied by the price people are willing to pay for that quantity. Forest Service arguments that it must "meet demand" ignore the fact that a reasonable and prudent seller will ensure that timber prices cover costs.

The crucial point is the way the Forest Service defines economically suitable timberland. Under NFMA, lands that are not physically or economically suited to timber management must be withdrawn by forest plans from timber production. The Forest Service defines "physically unsuitable" as lands that cannot be reforested in five years or that, if managed for timber, will suffer irreversible soil damage.[7] The number of acres of physically unsuitable lands is thus fixed by planners for each national forest.

Economically suitable lands, however, vary by alternative. The fact that timber makes or loses money when sold is irrelevant to whether it is economically suitable. Instead, lands are considered economically suitable if they are needed to meet the timber target of a particular alternative.[8] Each alternative may have different objectives, and so each alternative has a different number of economically suitable acres. The acres needed to meet the target of each alternative are considered suitable for that alternative.

Cost-Efficiency Versus Efficiency

The Forest Service maintains that all its alternatives are *cost-efficient*, but this does not mean that they are *efficient*. (The term *cost-effective* is sometimes used instead of *cost-efficient*; for the purposes of this book the two terms are identical.) Efficiency is attained only when present net value — the discounted value of all present and future benefits minus the discounted value of all present and future costs — is at its highest possible level. Money-losing timber sales are inefficient because they reduce present net value.

Cost-efficiency is the attainment of stated objectives, such as particular timber sale levels, at the least possible cost. Any timber sale level can be used as the objective, including one that is well above the efficient level. A high timber sale target could require that money-losing timber be sold, but it would be cost-efficient as long as it loses the least amount of money. Because most people do not understand the difference between efficiency and cost-efficiency, Forest Service claims that plans are cost-efficient lead many people to conclude that no timber sales will lose money. Others who know that sales lose money conclude that the Forest Service is simply falsifying data.

The FORPLAN computer program is used to produce cost-efficient alternatives. Planners enter the benefits and costs of timber sales into the computer and can ask the computer to propose the plan that produces the greatest possible present net value. For many forests that lose money on timber sales, such as the Gallatin (MT) and the Nantahala-Pisgah (NC) forests, FORPLAN proposes to cut no timber at all.[9] Planners, however, can tell the computer to produce any level of timber sales, and FORPLAN will do so at the least cost so long as the forest is physically capable of doing so.

Because RPA timber objectives were determined before any economic analyses were conducted, most forests include thousands of acres of money-losing timberlands in their suitable timber bases. For example, numerous studies have shown that nearly all timber sales in the San Juan National Forest lose money. The San Juan FORPLAN model in fact predicts huge economic losses for all alternatives considered in the San Juan environmental impact statement (EIS).[10] The EIS itself, however, gives no hint of this and instead claims that the benefits of all alternatives exceed their costs.[11] Only a close examination of internal planning documents reveals that these high benefits are based entirely on recreation values. The possibility that recreation values would be even greater in the absence of timber sales is completely ignored.

Plans that do not meet their RPA timber targets often base the timber objectives of their preferred alternatives on criteria just as dubious as the RPA objectives. Sometimes these criteria were an even more distorted version of the term *demand*. The Chugach Plan, for example, proposed a large increase in sales from that Alaska

Figure 4.2 Beaverhead FORPLAN Timber Values vs. Actual Prices

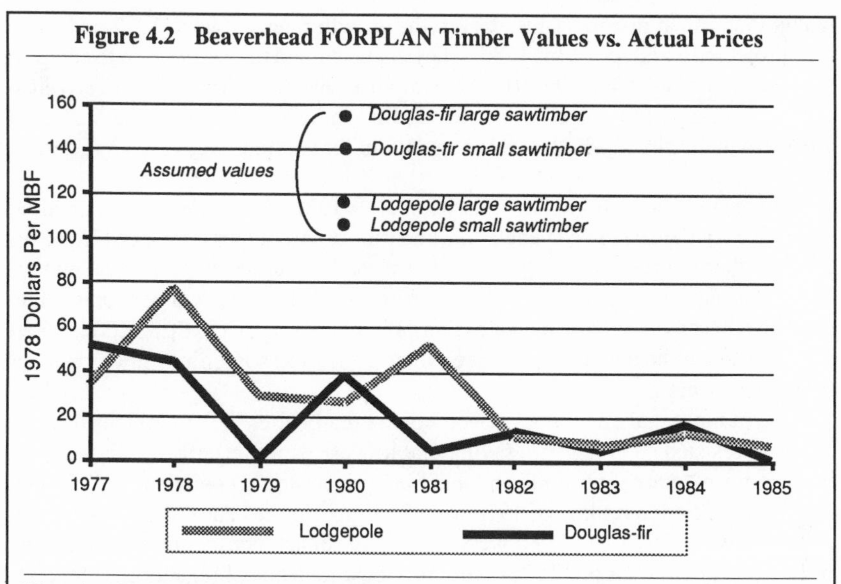

Historical high bids for Beaverhead National Forest timber have been much lower than the timber values assumed in the Beaverhead FORPLAN computer model. Values are in dollars per thousand board feet.

Source: CHEC, "Review of the Beaverhead Forest Plan and Final EIS" (Eugene, OR: CHEC, 1986), p .i.

forest in order to meet the supposed demand for timber. Forest planners calculated demand by taking two studies that estimated the capacity of local mills to process timber and adding them together.[12] The Forest Service completely overlooked the fact that this was double-counting the mills.

Plans for the San Juan, Chugach, and other forests included money-losing timberland in the timber bases simply because those lands were needed to meet timber objectives. Past timber sales from those lands had never collected more than a fraction of the costs of arranging the sales, and FORPLAN runs often indicated that proposed levels of timber management were causing a serious reduction in the economic value of the forests. Although these plans may be cost-efficient, they propose — often without public disclosure — to lose millions of dollars on timber.

Overestimated Timber Prices

Some plans used ingenious methods to fool not just the public but FORPLAN itself into believing that timber management is a sound economic proposition. The Beaverhead (MT) and Santa Fe (NM) forest plans, for example, assumed that timber values were higher than the forests had ever actually received in bid prices for their timber (figure 4.2).[13] When planning began in 1979, national forest timber had been selling at record prices, and there was little sign that this would not continue forever. Beaverhead planners assumed that prices would not only stay high but that they would increase at historic rates. Given the growth rates between 1972 and 1979, they projected a high price for 1980. They also assumed that prices would continue to grow after 1980. Since FORPLAN simplifies data by averaging timber values over a ten-year period, it used a value for the first ten years of the plans that was much greater than the already-high 1980 value.[14] Similar methods were used by Santa Fe planners.

Of course, prices did not rise; they fell after 1980. Many analysts blamed the high prices of the 1970s on speculation, and timber companies convinced Congress to allow them to return sales purchased at these high prices.[15] Given the historic quirk that planning began just before a major recession, many other forest plans were also based on overestimated timber prices. Average FORPLAN prices were typically twice the average sale prices since 1981, and sometimes prices reached five times as high.

Use of more realistic timber prices would greatly alter FORPLAN results. But planners resisted such realism, saying that low prices in the early 1980s were due to short-term "market turmoil" and that high prices would soon resume.[16] In fact, as a Forest Service review of 1973 to 1984 timber prices shows, the speculative prices between 1975 and 1980 were the extraordinary ones.[17] Prices have since returned to about their pre-1975 levels and show no signs of reaching the levels of the late 1970s for many years. Yet the prices bid between 1975 and 1980 governed the timber values used in most forest plans.

Rapidly Increasing Prices

As if to compound the problem of high timber prices, many plans also assumed that timber prices would rapidly increase over the next fifty years. Often this assumption was based on information developed for the 1980 RPA Program. The Forest Service asked Darius Adams, a forest economist from Oregon State University, and Richard Haynes, a forest economist with the Pacific Northwest Forest and Range Experiment Station, to build a complex model of the U.S. timber market to predict price *trends*, or the growth rate of future timber prices.[18] In 1979, the model indicated that Forest Service stumpage prices — the price of timber standing "on the stump" — would quadruple in the next fifty years.[19] These projections are often called the "1980 RPA trends."

Price trends are the modern expression of the belief held by many foresters that wood is essential for human existence. Wood substitutes, such as concrete, brick, and steel, are "energy intensive." Only food is more important than wood. As a Forest Service brochure on clearcutting once said, "We harvest timber because it is needed for man's survival."[20] Given this assumption, forest managers commonly assume that timber price increases of the future justify money-losing sales today. In fact, the opposite would be true. If prices are increasing as rapid as some forest plans claim, then the best decision would be to cut no timber as long as growth in prices exceeds the discount rate. This possibility was generally ignored by planners.

After assuming that timber prices today are higher than the forest has ever received in any year for its timber sales, Beaverhead planners went on to assume that prices will be growing at several percent per year for the next fifty years. Other Montana forest planners made more moderate assumptions about initial timber prices but used the same price trends. The Montana trends are far higher than those considered reasonable by any other economists — including the economists who prepared the model used to generate Forest Service trends. For example, figure 4.3 shows that prices in the Bitterroot National Forest are predicted to increase from about $100 to about $400 per thousand board feet.[21] In contrast, Adams and Haynes's 1980 price trends predicted prices would reach about $116 by the year 2030.[22]

High price trends allowed the Gallatin National Forest to predict that it would be making money in twenty or thirty years, but its relatively low initial price assumptions led to the conclusion that it would spend over $4.3 million per year (1987 dollars) more than it would collect on timber sales during the next ten years.[23] The Forest Service claims it needs to maintain the timber program for "community stability" and to ensure that mills will be around to consume the wood in the future when prices increase. Yet it would probably be cheaper for the Forest Service to buy the mills and pension off the workers than to continue these losses. If the Gallatin prediction is correct — and it may be an underestimate — taxpayers will lose over $61,000 per year for each of the 71 people whose jobs are supposed to depend on Gallatin National Forest timber sales.[24]

Late 1981 brought clear indications that the 1980 RPA price trends were wrong. The recession was more than the low end of the business cycle, while the timber boom at the end of the 1970s was something less than the high end. The boom was

Figure 4.3 Bitterroot Forest Prediction of Lodgepole Pine Values

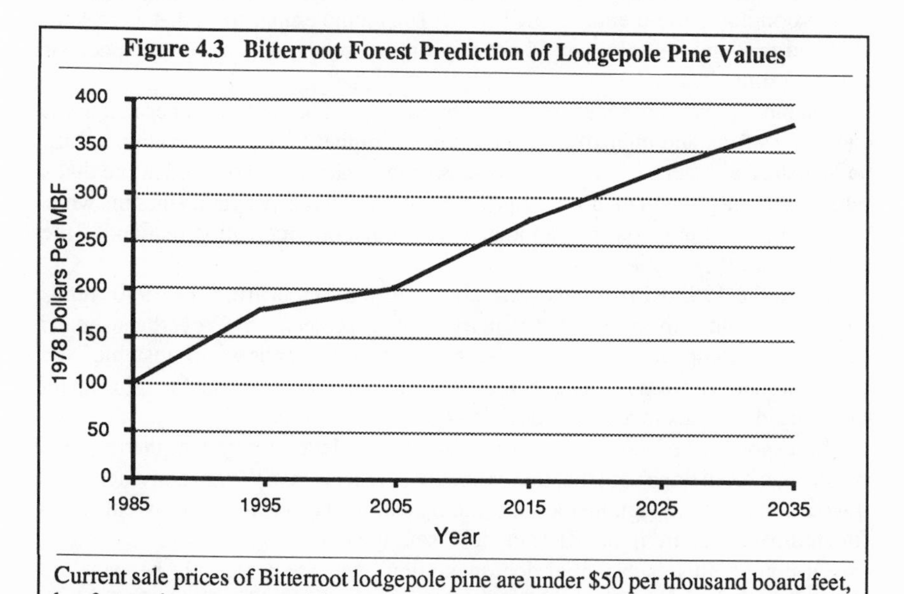

Current sale prices of Bitterroot lodgepole pine are under $50 per thousand board feet, but forest planners assumed that it would be worth nearly $400 per thousand board feet by 2035.

Source: CHEC, "Review of the Draft Bitterroot Forest Plan and EIS" (Eugene, OR: CHEC, 1985), p. 8.

caused by speculation due to poorly designed Forest Service sale procedures. The recession, meanwhile, was partly the result of major structural changes in the mortgage, banking, and housing industries — changes that would not likely be reversed. Any predictions made in 1979 were fairly suspect, and, in fact, most timber industry economists were far more pessimistic than the Forest Service.

In 1983, a revised Adams and Haynes model made new predictions for the 1985 RPA Program. These indicated that prices would rise but at a much slower rate than predicted in 1980.[25] This was still far more optimistic than most industry predictions, largely because the Forest Service required Adams and Haynes to assume that a "pent-up demand" for housing in the baby boom generation would produce a major increase in housing starts in the next few years.

Although the 1980 trends are discredited, most forest planners in Regions 1 (Montana and northern Idaho), 5 (California), and 8 (the South) continued to use them. Some planners in Region 3 (Arizona and New Mexico) shifted to the 1985 trends. Regions 2 (Colorado and eastern Wyoming) and 9 (the Midwest and Northeast) didn't use trends at all, and Region 4 (Utah, southern Idaho, and western Wyoming) stopped using trends in 1984.[26]

A Strong and Unjustifiable Bias

In 1983, it was still an open question whether the much-delayed Region 6 plans would use trends. Many Forest Service economists in Oregon and Washington opposed the use of trends and wrote a lengthy memo to the Washington office asking permission not to use them in FORPLAN. The memo emphasized that a "reasonable and prudent" investor would not rely on the trends, which were suspect for many reasons.[27]

Planners pointed out that the use of trends for timber but not other resources causes a "strong and unjustifiable bias" toward timber.[28] Since recreation values have historically been increasing as fast as timber values, "who could argue that a reasonable and prudent man would assume one level of real per capita income when projecting housing starts, and another when projecting recreation demand?" the memo asked.[29]

Since the 1985 RPA trends were completely different from the 1980 trends (figure 4.4), and both are very different from trends projected by Weyerhaeuser and other timber companies, the economists argued that the trends were "unstable" and of questionable accuracy. The memo suggested that no "reasonable and prudent man would base his investment decisions on such projections."[30]

In response, the Washington office sent a short letter noting that the region's memo was "well done" but that trends should still be used. "We are convinced that it would be an error to ignore the substantial documentation of long term real prices for stumpage that are in the RPA model," said the chief.[31]

Of course, this "substantial documentation" was available only because the Forest Service had contracted to have that information available. Similarly, the lack of information on cost trends or trends for other resources was due to the fact that the Forest Service had not contracted to have that information generated. The

Figure 4.4 RPA Program Timber Price Projections

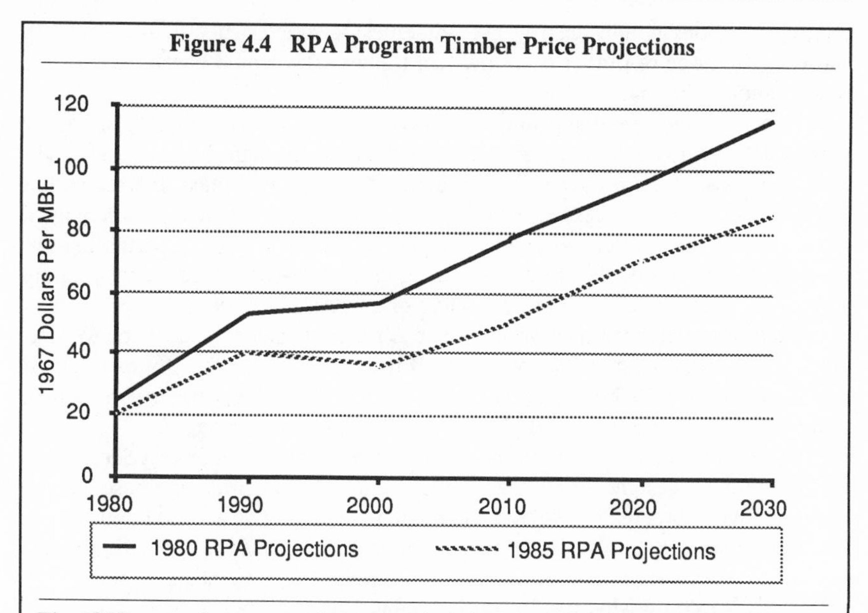

The 1980 projections are much more optimistic than the 1985 projections.
Source: Richard W. Haynes and Darius M. Adams, *Simulations of the Effects of Alternative Asssumptions on Demand-Supply Determinants on the Timber Situation in the United States* (Washington, DC: Forest Service, 1983), p. 14.

decision to contract with Adams and Haynes to generate timber price trends but not to hire someone to generate other resource trends represented a significant planning bias toward more timber sales.

The Region 6 memo overlooked another important argument against the use of price trends. Trends are needed in an economic analysis only when irreversible actions are being contemplated that trends might prevent. Instead, trends often led planners to take irreversible actions — such as proposing to build roads in roadless areas or to invade the habitat of endangered species — that would not have been proposed if trends were not used.

Anthony Fisher, an economist at the University of Maryland, and John Krutilla, of Resources for the Future, noted in 1974 that many environmental problems consisted of two alternative actions, one reversible and the other irreversible. The decision to harvest timber in a roadless area irreversibly destroys that area's value as wilderness, whereas the decision to postpone harvest for a few years does not irreversibly destroy its timber values. If the value of wilderness is known to be increasing, Fisher and Krutilla suggested that an "irreversibility premium," consisting of the present value of future wilderness use including price trends, be applied to the wilderness value.[32]

Suppose that a roadless area is available for backpacking, but several wilderness areas nearby are not being fully used by recreationists. The roadless area also has timber stands in it that could be sold if roads were built. However, the roads and logging activities would destroy the wilderness characteristics of the area. In addition, the price of timber is so low that timber sales would barely pay for the roads. Should the roads be built?

Projections indicate that primitive recreation use is growing rapidly. In a few years, all the nearby wilderness areas may be fully used, and the Forest Service will have to impose a permit system to regulate use. This will make the recreation opportunities in the roadless area much more valuable. Thus, it would be appropriate to use a trend for primitive recreation values to ensure that opportunities are not foreclosed that one day will produce high returns — and, when appropriately discounted, contribute to the present value of the roadless area.

A similar trend for timber would not be appropriate since the decision to postpone timber sales is generally reversible. If it were irreversible — for example, if there were a high probability that the timber might soon be destroyed by fire — then trends would be appropriate. But this would be an unusual case.

The decision not to build roads is completely reversible: The Forest Service can build roads to harvest the timber at any time. If future timber prices increase relative to road costs and recreation values, the roadless area may be able to contribute to the nation's timber supply. In the meantime, there is no need to second-guess the future and foreclose options by building roads today. Yet, as the next section will show, this is exactly what the Forest Service is doing.

Inflating the Timberland Base

Timber price trends in FORPLAN may foreclose options because FORPLAN determines the suitable timber base not by the number of acres needed to sustain the proposed first decade timber harvests but by the number of acres that will, if managed for timber, add to the economic value of the forest. Given high price trends, almost any forested acre, no matter how poor, would appear to eventually make money if its trees were sold.

Poor-quality acres, perhaps land in roadless areas that require expensive road construction, may be scheduled by FORPLAN for harvest in the year 2040. But they are included in the timber base immediately, so forest managers can begin road construction immediately. Thus, the price trends "inflate" the suitable timber base.

In many cases, trends in forest plans are higher than the Forest Service's 4 percent discount rate — especially when the trends were combined with the sometimes optimistic projections of future timber growth. If the trends and growth projections are accurate, this would mean that the present value of timber harvested in the future is greater than the value of timber harvested today. When this happens, FORPLAN will propose to delay timber harvests until the rate of increase of net timber values falls below the discount rate. This leads to some rather perverse results: Some FORPLAN models proposed to harvest no timber for fifty years, while others would harvest only a fraction of the sustained yield capacity.

To obtain what planners considered to be a reasonable level of timber harvests, planners in such cases were forced to require FORPLAN to harvest the desired volumes of timber. Unfortunately, planners failed to ensure that these minimum

Figure 4.5 Hypothetical Production Possibilities Frontier

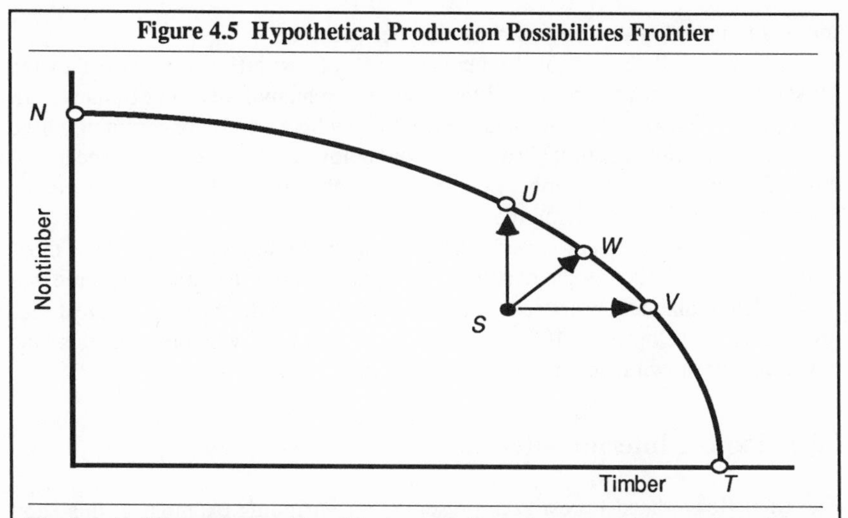

The frontier illustrates the trade-off between timber sale levels and the number of acres of land dedicated to nontimber uses. Any point on the frontier is cost-efficient, but a point below the frontier, such as *S*, is inefficient. By moving to *V*, timber can be increased without reducing nontimber, and by moving to *U*, nontimber acres can be increased without reducing timber. Both can be increased by moving to *W* or any point between *U* and *V*.

levels or *floors* approached the nondeclining flow level of the forests. Alabama forest planners used a floor of 75 million board feet in their preferred alternative and a floor of 45 million board feet in their wilderness alternative. The difference of 30 million board feet was blamed on wilderness, but in fact both alternatives could easily produce sustainable yields over 75 million board feet. The only difference was that planners chose a lower floor for the wilderness alternative.[33]

The trade-off between production levels of any two resources can be graphed as a *production possibilities frontier*. Figure 4.5 shows a hypothetical production possibilities frontier for timber and nontimber resources, the latter represented by the number of acres outside the suitable timber base. Zero timber production and all acres outside the timber base are represented by a point on the vertical axis. Maximum timber production, with all acres in the timber base, is represented by a point on the horizontal axis. The line connecting the two points represents the maximum amount of timber that can be produced given any number of acres in the timber base.

Without knowing the value of nontimber resources relative to timber prices, it is impossible to determine the efficient point on the frontier. However, all points on the frontier are cost-efficient, since all provide the maximum amount of one resource given any fixed amount of the other. Any point inside the frontier would not be cost-efficient. For example, point *S* is not cost-efficient because more timber can be produced without reducing nontimber outputs by moving to point *V*, while more land can be withdrawn for nontimber uses without reducing timber production by moving to point U — or any combination between.

Plans with inflated suitable timber bases are not cost-efficient because they fall far short of the production possibilities frontier. In many forests in Montana, Arizona, New Mexico, California, and the South — where trends are commonly used — the proposed level of timber sales was well below this frontier.[34] This means that either timber sales or the number of acres dedicated to nontimber use or both could be increased without reducing the other.

Thus, planners often proposed to build roads into roadless areas, such as those in Alabama, even though the destruction of this potential recreation resource was not needed to maintain proposed timber sale levels. The draft Gallatin Forest Plan proposed to develop some 100,000 acres of roadless land, when most of this land could be withdrawn from the suitable base without reducing timber sales.[35]

Selecting the Inferior Alternative

The final Beaverhead Forest Plan presented an intriguing example of this short sightedness. The selected alternative proposed to build roads into 137,000 acres of roadless land. Another alternative in the environmental impact statement proposed higher levels of timber harvest even though it would reserve all roadless areas from timber management. According to the EIS, this alternative was superior to the preferred alternative in almost every respect, including improved water quality and elk habitat.[36]

Two areas where the selected alternative appeared to be superior were total economic value and fish. Forest planners, however, confirmed that the fish projection was an error. And the selected alternative's high economic value was based on larger timber harvests in the future. Since the Beaverhead assumed that timber prices would rapidly increase, these timber sales greatly contributed to the economic value. When the price assumptions were corrected, however, the selected alternative proved inferior.[37]

Use of timber price trends frequently reverses the economic ranking of forest plan alternatives. By using price trends, high timber alternatives in sales-below-cost forests appear to have the highest economic value, mainly due to the high prices predicted for future sales. In reality, these alternatives have a lower value than alternatives that propose lower levels of timber sales. This happened in the Santa Fe, Deerlodge (MT), Beaverhead, and many other forests (figure 4.6).

Distortions from Road Misspecifications

In addition to overestimating timber values, many plans underestimated the costs of getting the timber — particularly road construction costs. A defect in early versions of FORPLAN is that road costs cannot be easily portrayed in the model. FORPLAN treats each acre separately, with no conception of the spatial relationship between acres. Thus, FORPLAN may propose to sell timber from a few acres in the center of a large roadless area and not recognize that many miles of new road

Figure 4.6 Alternative Deerlodge Forest Timber Values

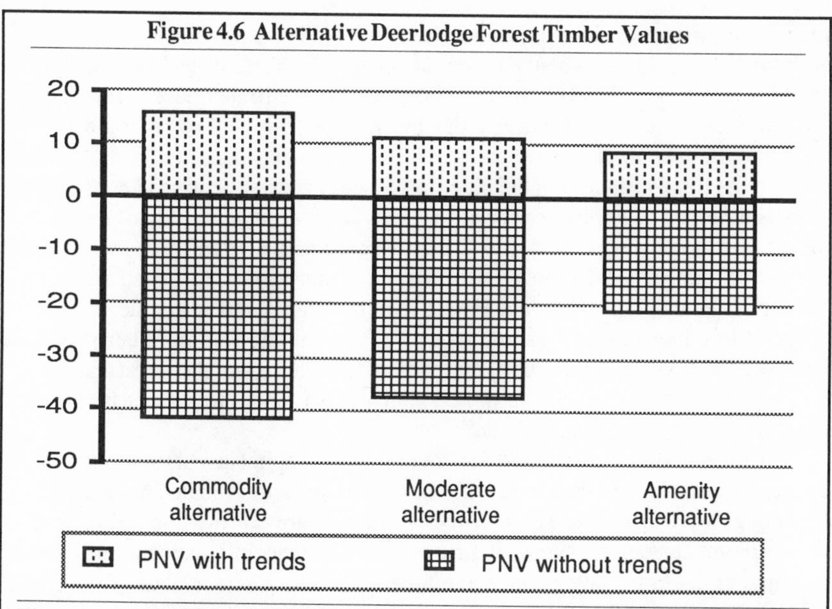

The economic ranking of alternatives changes when price trends are removed. Note also that the economic value is very negative without trends. Present net values (PNV) are in millions of dollars.

Source: CHEC, "Review of the Draft Deerlodge Forest Plan and EIS" (Eugene, OR: CHEC, 1985), p. 8.

construction will be needed to access those acres.

Most forest plans ignored this problem altogether. In fact, plans such as the Nantahala-Pisgah (NC), Carson (NM), and Nicolet (WI) simply assumed that road construction costs of harvesting timber in roadless areas would be identical to the costs of harvesting timber in areas with roads.[38] Although FORPLAN would normally tend to avoid roadless areas because of the high road costs, this tendency was eliminated by ignoring those costs.

In addition, to reduce discounted costs, FORPLAN would tend to minimize rates of road construction and stretch the length of time during which the road system would be built as long as possible — generally about 100 years. Planners considered this unrealistic. To meet a legal requirement that the size of clearcuts never exceed 40 to 60 acres, most planning teams estimated that the forest road system would have to be substantially complete in 50 years.[39] Since FORPLAN did not understand such problems, many planning teams simply let FORPLAN build the roads over 100 years. This effectively underestimated road costs and possibly led FORPLAN to include more acres in the suitable timber base. For example, a test run made by the Beaverhead Forest using more realistic road assumptions resulted in a reduction of the suitable timber base of 23 percent.[40]

Some forest planners attempted to overcome this problem by assuming that road costs would be higher in the first few decades and would decline to near zero after the fifth decade. Unfortunately, this had the unintended consequence that

FORPLAN simply decided to postpone as much road construction as possible until after the fifth decade — when roads could supposedly be built at a low cost.[41] With such low road costs, FORPLAN freely included in the timber base roadless lands whose timber values are negative today but will be positive if roads could ever be built into them at no cost.

Almost totally ignored in the planning process were several road construction models prepared by Forest Service economists and researchers. Malcolm Kirby, of the Pacific Southwest Forest and Range Experiment Station, prepared a transportation model called the Integrated Resource Planning model.[42] Fred Stewart, an economist at the Lolo National Forest (MT), wrote a much simpler computer program to compare the costs and returns of current and future timber sales in a watershed.[43] Ervin Schuster and Greg Jones, of the Intermountain Forest and Range Experiment Station, designed a more sophisticated model to predict future road construction and timber sale plans.[44]

All three models were designed to count roads as capital investments rather than operating costs. The models could thus be considered sophisticated methods of determining whether timber sales in roadless areas are worthwhile. Rather than simply count current-year costs against current-year receipts, as would be done in a simple cash flow analysis, or make unrealistic assumptions about future road values, as is done in the depreciation calculations of the *Timber Sale Program Information Reporting System* (*TSPIRS*), these models measure the discounted value of all timber to be accessed by a road network against the discounted cost of the roads.

The Stewart and Schuster and Jones models were tested on selected timber sales in western Montana national forests. Schuster and Jones found that, when 1980 RPA timber price trends were used, returns on future timber sales may pay for investments in roads.[45] However, tests of Stewart's model found some instances where even the 1980 price trends did not raise benefits above costs.[46] Moreover, when no price trends were used, road costs were greater than timber values in all tests.

The use of price trends is analytically questionable for the reasons already described. Otherwise, these models represent sound procedures for estimating the net values of timber in roadless areas. Yet they have rarely, if ever, been applied in forest planning.

Ironically, the Schuster and Jones model was funded, at least in part, by the congressional appropriation for developing a timber cost accounting system.[47] Yet, as described in chapter 2, the authors of the *TSPIRS* discarded the model in favor of the controversial method of road depreciation.

Roads are one of the most important costs of timber management, yet most forest planners have failed to develop or use any system that realistically compares road costs with timber values. As a result, Forest Service computer models almost invariably underestimate the true costs of road construction.

Making the Analyses Realistic

Despite all the problems with forest planning, many Forest Service economists defend the process. Clark Row, an economist who until recently was director of the Forest Service's Office of Policy Analysis, claims that economic efficiency is "the single most important decision criterion for all long-term planning decisions." Row insists that the problems with the planning process are actually "improvements in economic methodology." In particular, Row calls proposals to use discount rates over 4 percent a "major policy problem." "Almost no investments in timber growing in the United States — public or private — return [10 percent] on investment," claims Row. "Trees just don't grow that fast." Using the 4 percent rate, says Row, along with "many other improvements in economic methodology, the Forest Service is now finding that economic efficiency analyses are, in their managers' judgment, more realistic."[48]

Row does not describe the "many other improvements in economic methodology" in detail. Presumably, they include the use of cost-efficiency instead of efficiency, the failure to calculate soil expectation value, the counting of reforestation costs against the value of timber harvested today rather than as an investment in future timber growth, the assumption that timber values are rapidly increasing, and the definition of demand as a point rather than as a line. All of these "improvements" make Forest Service planning "more realistic" in the judgment of national forest managers — that is, they make the answers agree with the managers' preconceived notions. Not surprisingly, many forest plans appear to be little more than justification documents confirming the status quo.

Economic efficiency has not been a major motive for national forest planning or management. Instead of using economic tools to identify the optimal method of managing the national forests, the designers of the process and individual forest planning teams have deliberately or inadvertently biased plans to justify timber production and cover up the frequent losses of the timber program. The result is that forest plans mislead the public into believing that timber management is an efficient use of the national forests when, in most cases, it is extremely wasteful.

Notes

1. William F. Hyde, "Timber Economics in the Rockies: Efficiency and Management Options," *Land Economics* 57(4):630–637.
2. Randal O'Toole, *The Citizens' Guide to FORPLAN* (Eugene, OR: CHEC, 1983), 56 pp.
3. CHEC, "Review of the Draft Shoshone Forest Plan and EIS" (Eugene, OR: CHEC, 1985), p. 12.
4. Alan McQuillan, "The Declining Even Flow Effect — Non Sequitur of National Forest Planning," *Forest Science* 32(4):960–972.
5. "Departure Analysis," memo from Klamath Forest supervisor to regional forester on file at Klamath Forest supervisor's office, Yreka, CA, 29 June 1982, p. 1.
6. USDA Forest Service, *San Juan National Forest Plan Final EIS* (Durango, CO: Forest Service, 1983), p. II-34.
7. 36 *Code of Federal Regulations* 219.14(a).
8. 36 *Code of Federal Regulations* 219.14(c).

9. USDA Forest Service, *Gallatin National Forest Plan Draft EIS* (Bozeman, MT: Forest Service, 1985), p. B-65.
10. CHEC, "Economic Analysis of the San Juan Forest Plan" (Eugene, OR: CHEC, 1983), p. 4.
11. USDA Forest Service, *San Juan National Forest Plan Final Environmental Impact Statement* (Durango, CO: Forest Service, 1983), p. II-39.
12. CHEC, "Review of the Final Chugach National Forest Plan" (Eugene, OR: CHEC, 1984), p. 14.
13. CHEC, "Review of the Beaverhead National Forest Draft Plan and EIS" (Eugene, OR: CHEC, 1983), pp. 2–3; "Affadavit of Randal O'Toole on Santa Fe Forest Plan" (Eugene, OR: CHEC, 1983), p. 7.
14. CHEC, "Review of the Draft Beaverhead Plan," p. 5.
15. Timber Contract Modification Act of 1984.
16. USDA Forest Service, "Responses to Comments Related to Forest Plans in R5," memo on file at Region 5 regional office, San Francisco, CA, 1986, p. 24.
17. Florence K. Ruderman and Richard W. Haynes, *Volume and Average Stumpage Price of Selected Species on the National Forests of the Pacific Northwest Region, 1973 to 1984* (Portland, OR: Forest Service, 1986), pp. 4–7.
18. Darius M. Adams and Richard W. Haynes, *The 1980 Softwood Timber Assessment Market Model: Structure, Projections, and Policy Simulations* (Washington, DC: Society of American Foresters, 1980), pp. 9–38.
19. Ibid., p. 51.
20. USDA Forest Service, *Patience and Patchcuts* (San Francisco, CA: Forest Service, 1974), p. 17.
21. CHEC, "Review of the Bitterroot National Forest Draft Plan and EIS" (Eugene, OR: CHEC, 1985), p. 4.
22. Adams and Haynes, *Softwood Timber Assessment*, p. 14.
23. USDA Forest Service, *Gallatin National Forest Plan Draft Environmental Impact Statement* (Bozeman, MT: Forest Service, 1985), pp. II-110–111.
24. CHEC, "Review of the Gallatin National Forest Plan and EIS" (Eugene, OR: CHEC, 1985), p. 21.
25. USDA Forest Service, *1985-2030 Resources Planning Act Program Final Environmental Impact Statement* (Washington, DC: Forest Service, 1986), p. F-16.
26. "Region-by-Region Overview of Forest Planning," *Forest Watch* 6(10):20–29.
27. "Timber Value Trends in Forest Planning," unpublished paper on file at Forest Service Region 6 office, Portland, OR, 23 November 1983, p. 4.
28. Ibid., p. 2.
29. Ibid., p. 2.
30. Ibid., p. 4.
31. "Real Price Trends for Timber," unpublished memo from Washington office to Region 6 on file at the Portland office of the Forest Service, 16 January 1984, 1 p.
32. Anthony C. Fisher and John V. Krutilla, "Valuing Long Run Ecological Consequences and Irreversibilities," *Journal of Environmental Economics and Management* 1:96–108.
33. CHEC, "Review of the Forest Plan and EIS for the Alabama National Forests" (Eugene, OR: CHEC, 1985), pp. 8–11.
34. CHEC, "Review of the Four Northern Arizona Forest Plans" (Eugene, OR: CHEC, 1986), pp. 27–29.
35. USDA Forest Service, *Gallatin Draft EIS*, p. II-103.
36. CHEC, "Review of the Beaverhead Forest Plan and Final EIS" (Eugene, OR: CHEC, 1986), p. 10.
37. Ibid., p. 11.
38. CHEC, "Review of the Nantahala-Pisgah National Forests Draft Plan and EIS"

(Eugene, OR: CHEC, 1984), p. 2; CHEC, "Review of the Draft Carson Forest Plan and EIS" (Eugene, OR: CHEC, 1985), p. 2; CHEC, "Review of the Nicolet National Forest Draft Plan and EIS" (Eugene, OR: CHEC, 1985), p. 12.

39. USDA Forest Service, *Bridger-Teton National Forest Plan Draft Environmental Impact Statement*, p. B-VI-3.
40. CHEC, "Review of the Final Beaverhead Plan," p. 5.
41. CHEC, "Review of the Bitterroot Plan," pp. 7–9.
42. Malcolm W. Kirby, et al., *Guide to the Integrated Resource Planning Model* (Berkeley, CA: Forest Service, 1980), pp. 2–3.
43. USDA Forest Service, "Multiple-Use Timber Sale Examples" (Washington, DC: Forest Service, 1985), pp. 4–6.
44. Ervin G. Schuster and J. Greg Jones, *Below-Cost Timber Sales: Analysis of a Forest Policy Issue* (Ogden, UT: Forest Service, 1985), p. 5.
45. Ibid., p. 11.
46. Fred Stewart, "Assumptions for Tolan Creek Economic Analysis," unpublished memo on file at the Lolo National Forest, Missoula, MT, 29 pp.
47. "Forest Service Studies and Actions Related to the Economics of the Timber Sale Program," unpublished paper on file with the Forest Service, Washington, DC, 14 February 1986, p. 8.
48. Clark Row, "Forest Service Budget Maximization: A Dissent." *Renewable Resources Journal* 2(4):5–7.

Chapter Five
Multiple-Use Clearcuts

"By the way," says Iacocca to the chief one day, "I've been looking over your report to the board, and I can't find any reference to finance charges on the money we have borrowed in the past or future."

"We didn't include past finance charges at all because finance charges are contrary to the sustained production of automobiles," says the chief. "However, we did discount future revenues and costs using a 4 percent discount rate."

"Where are we supposed to get loans at 4 percent?" asks Iacocca.

"I am sure your bankers will give you loans at that rate," answers the chief. "Just tell them how many jobs you will be creating when you increase auto production. You can also tell them that your new cars will make people happier, increase productivity, and improve the quality of life."

"Why will the bankers loan us money for those things?" asks Iacocca.

"I don't know, but I'm sure you can figure out a way to convince them. The Forest Service was always able to do so."

According to this imaginary chief, automobile production carries a number of *side benefits*, including employment, the enjoyment of car ownership, and the increased productivity of the companies using cars and trucks. The chief expects that bankers should be willing to subsidize these side benefits, just as the Forest Service expects all taxpayers to subsidize national forest management even though only some of the taxpayers receive the direct benefits of management. However, it is not clear that the side benefits the chief mentions really exist.

The Forest Service's 1984 white paper on the role of below-cost timber sales in forest management notes that the "National Forests are managed for numerous 'products' and amenities in addition to timber. But many of these additional benefits are derived, at least in part, through the process of harvesting timber."[1]

Specific side benefits of timber management mentioned in the paper include improvement of "wildlife habitat sites or watersheds" and protection of the forests from insects and disease. The paper also notes that roads "make possible protection of the forests from fire" and "enhance recreation, wildlife management, grazing, and other uses of the forest." The paper notes that all these benefits are considered in the forest plans. "The social, environmental and economic effects of all management options (including timber, forage, wildlife, water, outdoor recreation, wilderness, and minerals) are analyzed on a Forest-wide basis" in these plans.[2]

Many forest plans use nontimber resources to justify money-losing timber sales.

The plans sometimes claim that roads increase recreation opportunities. Timber cutting is supposed to improve wildlife habitat. Clearcutting increases water production, which is claimed to have a high value in the arid West. Roads and timber cutting are also supposed to reduce fires, insects, and disease. In addition, the Forest Service often says that timber management is needed to protect local community stability. However, a close examination of these claims reveals that they are rarely valid.

"I Did As I Was Told"

One frequent claim is that timber-related roads benefit recreationists. For example, a sixteen-page glossy color brochure titled *National Forest Roads for All Uses* was distributed with every forest plan published by the Eastern Region. "Multiple use management needs roads," says the brochure, which adds that roads make more recreation opportunities available, "resulting in increased public benefit."[3] This brochure was apparently designed to reduce opposition to proposed road construction programs.

Most eastern national forests include several roadless areas inventoried during the Forest Service's roadless area review and evaluations. Yet many Eastern Region plans essentially denied that such areas existed. For example, the environmental impact statement for the Chequamegon (WI) Plan noted that all roadless areas not designated wilderness would be managed for "multiple use."[4] Elsewhere

Figure 5.1 Bitterroot Forest Recreation Use and Capacity

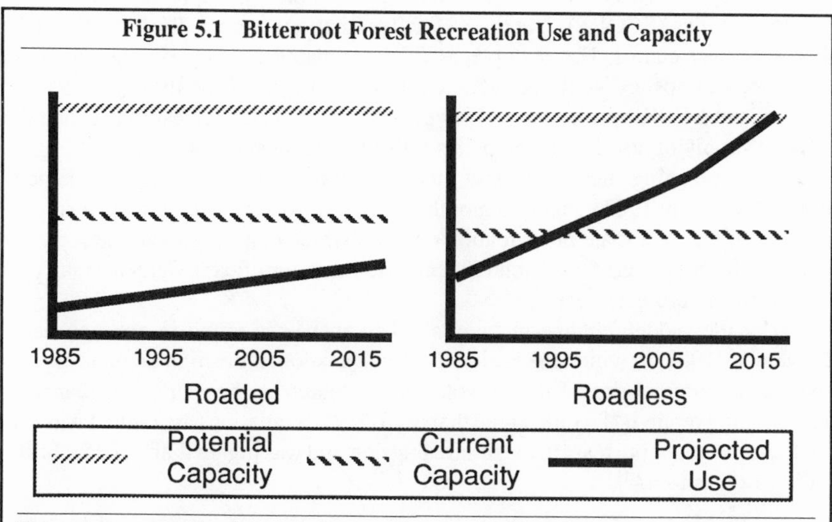

The existing capacity for roaded recreation on the Bitterroot — and most other — forests is far greater than the current or projected use, while roadless recreation will soon be in short supply.

Source: USDA Forest Service, *Bitterroot National Forest Plan Draft EIS* (Hamilton, MT: Forest Service, 1985), p. II-5.

in the EIS, "multiple use" was defined as requiring 4.5 miles of road per square mile of land. This definition was obtained, said the EIS, by estimating the number of miles needed for timber sales.[5] Most Chequamegon Forest sales lose money, and primitive recreation opportunities are in very short supply, a fact that applies to most Eastern Region forests. Yet no consideration was given to managing these areas for primitive recreation.[6]

Most forest planning teams have concluded that the existing road network is adequate for current and projected recreation use. For example, recreationists on the Bitterroot National Forest, which is nearly three-quarters roadless, might be thought to benefit from new road construction. Yet planners indicate that the current road system is more than enough to satisfy recreation use that requires roads until well past 2030 (figure 5.1).[7] Meanwhile, roadless recreation use is expected to exceed capacity in as short a time as fifteen years.[8] Recreationists who use roads would probably not benefit from more roads. Indeed, road construction might reduce roadless recreation values, particularly if existing wilderness areas are expected to become crowded in the near future. As Resources for the Future economist John Krutilla points out, "Where the new facilities would fail to provide qualitatively better recreation, and excess capacity existed in the roaded area of the forest, no additional net recreation benefits would result from building a new access road."[9]

The Hoosier National Forest (IN), which is in the Eastern Region, may be the only national forest whose analysts claimed that recreation demand required new road construction. Although the forest is extensively roaded, planners projected that the demand for road-related recreation is far greater than the current capacity. This claim was based on information in the Indiana State Comprehensive Outdoor Recreation Plan (SCORP). Yet the SCORP predictions were in direct conflict with Forest Service claims. For example, SCORP estimated that the existing number of developed campsites would be sufficient to meet projected use through 1995, the limit of the SCORP projections. However, the Forest Service estimated that the current supply of developed campsites falls 90 percent short of demand.[10]

Forest planning files reveal that the recreation planner had prepared demand estimates in July 1982. Just two months before, however, he had made a significantly different estimate of recreation demand which concluded that the forest's capacity for all kinds of road-related recreation was significantly greater than current or future projected use.

Why the sudden change in figures? The answer is found in a memo dated November 1982, in which the regional office questioned the high demand figures used in an internal draft EIS. A note on the memo in the recreation planner's handwriting replied, "I would agree that my 7/7/82 calculations are high! I was told by the forest planning team to make sure that demand was higher than our capability. I did as I was told."[11]

Clearcuts for Wildlife

While few planners are willing to fabricate recreation data to justify timber cutting, many are willing to use wildlife research selectively to support increased timber sale levels. A well-documented research project in northeastern Oregon indicated that deer and elk thrive best when the ratio of cover to forage is about 40 to 60. In other words, ideal big game habitat in northeastern Oregon consists of a mosaic of patches of heavy timber, which animals can use for shelter from predators and temperature extremes, and patches of open areas, which provide edible vegetation. The timber patches should occupy about 40 percent of the area, and the forage patches about 60 percent.[12]

At first glance, this seems like a strong argument for timber cutting. Although the researchers warned that their results should be applied only to the Blue Mountains of northeastern Oregon and southeastern Washington, forest planners in such diverse areas as New Mexico and North Carolina have used this research to justify money-losing timber sales.[13]

In fact, planners have been misreading — or selectively reading — the research. The Blue Mountain biologists defined "cover" as forest stands with at least 70 percent crown closure. In other words, when viewed from the air, no more than 30 percent of the ground is visible; the remainder is covered by trees. Stands with less than 70 percent crown closure qualify as "forage" areas because the shrubs in the openings provide food, while the lack of dense shelter fails to provide protection from predators or temperature extremes.[14]

Nearly every forest plan in New Mexico claimed high benefits to wildlife from the openings created by timber cutting and the "improvement" in the cover-forage ratio. Yet many, if not most, New Mexico forests have less than 70 percent crown closure. Timber cutting will turn forests that now have less than 40 percent cover into forests with almost no cover.[15]

Planners also selectively apply the 40-60 cover-forage ratio. In the Gallatin Forest supervisor's office, I reviewed some aerial photos of the forest with Matthew Reid, a local wildlife biologist. Reid pointed to an area of patchy timber. "This area is perfect elk summer range," he said, "with both open areas for forage and dense timber for cover. I've seen large numbers of elk occupy the area every summer." A map of the forest plan indicated that the area he referred to was slated for intensive timber management, with no concern for wildlife. "Any timber cutting in this area will reduce the cover and eliminate its value as elk habitat," said Reid.

A nearby area was proposed for "timber-wildlife" management — a prescription designed to maximize the area's value for wildlife. The aerial photo showed that this area included dense stands of timber with few openings. "Timber sales in this area may create forage and increase the wildlife value," noted Reid, "but there would be no need to do so if the existing summer range nearby were not decimated by sales." Timber management on the Gallatin Forest loses $4 million per year, and one of the the main justifications is that it will improve habitat for deer and elk. But, according to Reid, there would be no need to "improve" elk habitat if it were not destroyed by timber sales in the first place.

Conflicts between game and timber cutting are compounded in the northern Rocky Mountains by the spread of spotted knapweed, an exotic plant that is poisonous to deer and elk. Uninfested areas frequently decline in habitat quality after the weed is introduced, primarily through new road construction. Although hunting values are greater than timber values in forests like the Beaverhead and Gallatin, Forest Service plans call for extensive road construction and timber cutting in hundreds of thousands of acres that are now considered prime game habitat.[16]

The consequences of errors in these claims might not be significant, since deer and elk are unlikely to be driven to extinction by clearcutting. Of greater concern is the Forest Service's treatment of rare or endangered species.

Grizzly-Timber Emphasis

The grizzly bear is close to extirpation in the vicinity of Yellowstone Park. The greatest threat to the bear is human-caused deaths, often the result of encounters between humans and bears in developed areas. Biologists therefore recommend avoiding activities that might lead to more encounters between bears and humans.

The Gallatin National Forest borders Yellowstone Park on the north and west and is frequently used by grizzlies who roam beyond the boundaries of the park. The draft Gallatin Plan allocates over 108,000 acres of forestland that is described as "prime grizzly bear habitat" to a prescription called "grizzly-timber emphasis." The plan says that timber in this area will be harvested to enhance grizzly habitat.[17]

In reality, planners have no idea how timber management might be able to enhance grizzly bear habitat. They say that the grizzly feeds on the seed from the whitebark pine and speculate that timber management might be able to promote production of this tree. But researchers are not familiar with the successional relationships of whitebark pine and other species. If the pine is a climax species — able to reproduce itself without outside disturbances — no timber management would be needed to maintain it.

Despite this lack of information, the draft plan assumed that all 108,000 acres in the grizzly-timber emphasis area would produce the maximum possible amount of timber.[18] No allowances were made for reduced timber yields due to possible constraints to protect the grizzly bear. Yet roads constructed to provide access to timber in this zone are likely to invite human-grizzly encounters that may lead to additional grizzly deaths. Although the Forest Service says it will close at least some of the roads in the grizzly-timber zone, a study of road closures in prime grizzly habitat on the Flathead National Forest found that 38 percent of the closures were ineffective.[19]

Thus, in order to provide the dubious benefit of pine seed to the grizzly — a benefit the Forest Service does not even know it can provide — the Gallatin Plan seriously risks hampering the efforts to recover the Yellowstone bear population.

Questionable benefits to the grizzly bear from timber cutting were also claimed by the Flathead National Forest, which provides habitat for the Glacier Park population of grizzlies. The forest assumed that roads would reduce the number of bears by 20 percent, but that a decision not to build roads would increase grizzly numbers by 20 percent. Planners also assumed that clearcuts would increase vegetation diversity, which would also increase the grizzly population by 20 percent. A management scheme was devised for cutting without roads, which planners claimed would increase bear numbers by 40 percent.[20]

In reality, it is unlikely that a decision not to build roads into a roadless area will inspire grizzlies to increase their numbers. The idea that clearcutting increases diversity is also questionable, yet it is claimed in nearly every forest plan.

Biological Deserts

Species that are dependent on old-growth forests represent another conflict with timber. Foresters traditionally use the term *old-growth* for all forests older than rotation age. Since rotation ages in the national forests are usually between 50 and 120 years, any timber stands older than about age 100 are likely to be called old-growth. Early wildlife researchers examining such old-growth stands found them to be extremely dense, with closed tree canopies that captured all the sunlight. With little vegetation at ground level other than the trunks of trees, few species of wildlife could find a home. Thus, old-growth forests were frequently referred to as "biological deserts" that were incapable of supporting any life except the trees themselves. Such claims went unchallenged through the early 1970s.

Then, in 1976, a graduate student at Oregon State University named Eric Forsman studied a little-known bird called the northern spotted owl and discovered that it depended almost entirely on old-growth Douglas-fir forests.[21] While individual birds could survive in second-growth forests, the species appeared to need very old stands of trees — at least 250 years of age or more — for breeding. This was partly due to the fact that the principal prey species of the owl — the northern flying squirrel — also lived almost exclusively in old-growth.

In addition, the great horned owl tended to prey on spotted owl young. The larger owl swoops down on spotted owl nests, knocks eggs or juveniles from the tree, then feeds on them on the forest floor. Spotted owls could protect their young by building their nests in broken-topped trees. The side branches of such trees tend to turn upward to form new tree tops, protecting spotted owl nests from the great horned owl. Large broken-top trees are found mainly in forests with trees older than 250 years.

After Forsman published his thesis, other researchers found many other wildlife species that are dependent on very old forests. At latest count, some thirty species of birds, mammals, and other vertebrates appear to rely on older Douglas-fir forests.[22]

The pileated woodpecker, the largest surviving woodpecker in North America, carves holes in large snags for its nests.[23] Logging the last ponderosa pine and Douglas-fir old-growth would cause this species to follow its eastern cousin, the ivory-billed woodpecker, into extinction. The thick, craggy bark of old-growth Douglas-fir provides a home for the long-eared myotis, an increasingly rare species of bat. In southeast Alaska, the Sitka black-tailed deer relies on old-growth for food and shelter in winter snows. According to Forest Service and Alaska State researchers, "The amount of old-growth, critical winter range is likely to directly determine the carrying capacity for deer" where there are heavy snowfalls.[24]

Biologists came to divide stands that foresters called old-growth into two types: "mature" timber, including stands between 100 and 250 years old, and "old-growth," including stands over 250 years old. The "biological deserts" of lore were actually the immature and young mature forests between 50 and 150 years of age.[25] Of course, exact ages depend on the species and the forest site.

Clearcuts in such biological deserts seem to benefit wildlife because they create openings in which grasses, shrubs, and other vegetation can grow near ground level, available to deer and other browsers and foragers. Such vegetation is often already present in true old-growth forests, since these forests usually have natural openings through which light can pass. When each tree consumes 2 to 5 percent of an acre, the death of one tree can create a significant opening. Thus, truly old-growth forests are often characterized by multiple layers: the top layer being the oldest trees, a second layer composed of younger trees, and a third layer of seedlings and shrubs.

Old-growth forests also have many dead trees (snags) of various sizes and states of decomposition, as well as fallen logs, which can be very large. An old-growth tree falling across a stream is much more likely than a younger tree to form a small dam, which slows the water and allows sediment to drop out. As a result, old-growth forests produce the cleanest water by far.

In the Pacific Northwest, old-growth timber is highly valued for wood. Protection of large expanses of old-growth as habitat for species such as the spotted owl may be difficult to justify on a strictly economic basis. But through the Endangered Species Act, Congress has suggested that the preservation of viable populations of all wildlife species is an appropriate social goal.

However, the question here is not how much valuable old-growth timber should be withdrawn from management to protect rare species of wildlife, but rather why forest plans with money-losing sales claim that such sales are needed to enhance wildlife habitat. These claims are based on a narrow view of forest diversity.

The ecological characteristics of old-growth forests — multilayered canopies, snags, fallen logs, and clean water — ensure that old-growth forests provide a diversity of wildlife habitat unmatched by timber of any other age. This sort of diversity might be called *vertical* diversity. A regulated forest provides another sort of diversity, which might be called *horizontal* diversity. The diversity within any given stand of trees is low. But if the clearcuts (and thus stands) are small enough, travel between stands reveals a high level of diversity. This diversity can be deceptive, however, since a regulated forest will have few stands of trees older than about 100 years. Animals and plants that rely on large snags, broken-top trees, thick, craggy bark, or other characteristics of old-growth forests will find survival difficult in a managed forest.

Many forest plans use only a horizontal scale to measure diversity. Planners in Arizona, for example, defined "an acceptable diversity as a diversity of cover types and age classes of all vegetation types at any given point of time."[26] Based on this definition, they estimated that the highest diversity forest will include 3 percent old-growth and fixed percentages of other age classes. Planners divided forests into blocks of about 10,000 acres and found that many of these blocks had more than 300 acres, or 3 percent, in old-growth stands. Thus, to "improve" diversity, the plans called for liquidation of the "surplus" old-growth. Timber sales in many Arizona forests lost money but were justified on the grounds that they improved wildlife habitat.

The quality of wildlife habitat in a second-growth forest is reduced even further if the forest is intensively managed, with frequent thinnings during each rotation. Thinnings reduce the quality of cover provided for game species such as deer and elk, making it more difficult for them to hide from hunters and predators and to find protection from temperature extremes.

Biologists say that such timber management "simplifies" and "homogenizes" the forest. According to forest ecologist Jerry Franklin, "A universal application of such an approach is unwise, since it reduces the options of managers and of society. It also ignores major elements of the ecosystem, something done only at peril."[27] Where Franklin recommends modifications of management in the valuable coastal Douglas-fir forests to minimize simplification, there is no need for simplification, no need for timber management at all, in most Rocky Mountain and Intermountain forests.

Some species seem to require not simply a few older trees but large, unfragmented expanses of undisturbed forest. Elk, for example, shun heavily traveled roads.[28] The spotted owl seems to be losing ground to a species common in the Rocky Mountains but new to the Pacific Northwest, the barred owl. While the barred owl seems to require old-growth forests, it does not need as large acreages as the spotted owl. This is partly due to the fact that female barred owls are slightly larger than males, so the sexes feed on different animals. Male and female spotted owls are about the same size, so a breeding pair requires a larger acreage of old-growth to find enough prey. As the size of remaining old-growth stands dwindles, the barred owl appears to be occupying habitat once used by the spotted owl.[29]

Despite a growing understanding of the value of old-growth to wildlife, many forest plans continue to claim that intensive timber management provides better wildlife habitat than unmanaged forests. This may be true for some species, but given that most forests are scheduled for management, wildlife species that are counted as "rare" are the least likely to find managed forests to be suitable habitat.[30]

Forest Planning in 1776

Even where old-growth is considered an important component of the forest, managers of forests that lose money on timber continue to insist that regular harvests of old-growth are necessary for wildlife. Rather than protect the remaining stands of old-growth, complicated schemes are developed for growing new old-growth. Such plans are being made in the Gifford Pinchot Forest in Washington and in several California forests.

Yet it is ecologically, politically, and economically dubious whether new old-growth can be grown. Biologists admit that they don't fully understand the processes that created existing old-growth forests or which components of old-growth are really important, and many suggest that cutting existing old-growth should be considered irreversible.[31]

The political problems were succinctly raised by Olof Wallmo, a biologist with the Alaska Department of Fish and Game, who suggested that "planning to grow trees on a 200-year rotation is like participating in the First Continental Congress on the forest planning committee."[32] After all, a stand of trees on public land that is scheduled to become old-growth is potentially merchantable for nearly 200 years before it actually becomes old-growth. It would only take one administrator like James Watt during that time to decide to harvest those trees — and who would defend trees that won't become old-growth for several generations?

These considerations are reflected by an economic analysis of old-growth. Managing old-growth using long rotations requires many more acres of land than simply protecting existing old-growth. For example, suppose that 300-year rotations are to be used, and the stands would be considered old-growth between the ages of 250 and 300. Then 6 acres of land would need to be so managed for every acre that would be old-growth at any given time.

Perhaps more importantly, the cost of waiting for trees to become old-growth is extremely high. Even using the Forest Service's 4 percent discount rate, an acre of 300-year old timber would have to be worth nearly $13 million to make it desirable to spend $100 to establish it today. This is simply another way of saying that someone—James Watt's great-grandson, say, who becomes chief of the Forest Service in 2088—is likely to cut the trees before they actually become old-growth. Who would protest the cutting of 100-year-old trees simply because they might become old-growth in another 200 years? It is both less expensive and more politically reliable to simply preserve existing stands of old-growth.

Old-growth protection may be expensive in terms of the lost timber values in those forests that have valuable timber, but not in forests that traditionally lose money on timber sales. Thus, it is especially disturbing to see forests such as the Kaibab (AZ) and Lewis and Clark (MT) propose to harvest their remaining stands of old-growth at an enormous cost to taxpayers in the name of improving wildlife habitat.[33]

Wildlife is integrated into forest planning through the designation of a number of species as "indicator" species. Each species represents a particular ecotype or environmental condition. The spotted owl is an indicator of old-growth, so most Pacific Coast forests use the spotted owl as one of their indicator species. Because the owl uses a large territory, planners assume that protection of the owl will also protect other old-growth-dependent species, such as the silver-haired bat and the northern flying squirrel. Yet the plans frequently fail to ensure that indicator species will be protected.

The spotted owl is such an important issue that most California forests have incorporated spotted owl "yield tables" into their FORPLAN computer models. These tables purport to track owl habitat quality according to the age of timber, so that FORPLAN can predict the number of spotted owls produced at various levels of timber harvesting. But the spotted owl yield tables are skewed to actually encourage timber harvesting and rapid liquidation of old-growth.

In the Sierra National Forest, the owl is projected to find suitable habitat in any timber older than 60 years. Optimum habitat is supposed to be found in timber between 120 and 230 years. Timber older than 230 years declines in habitat quality.

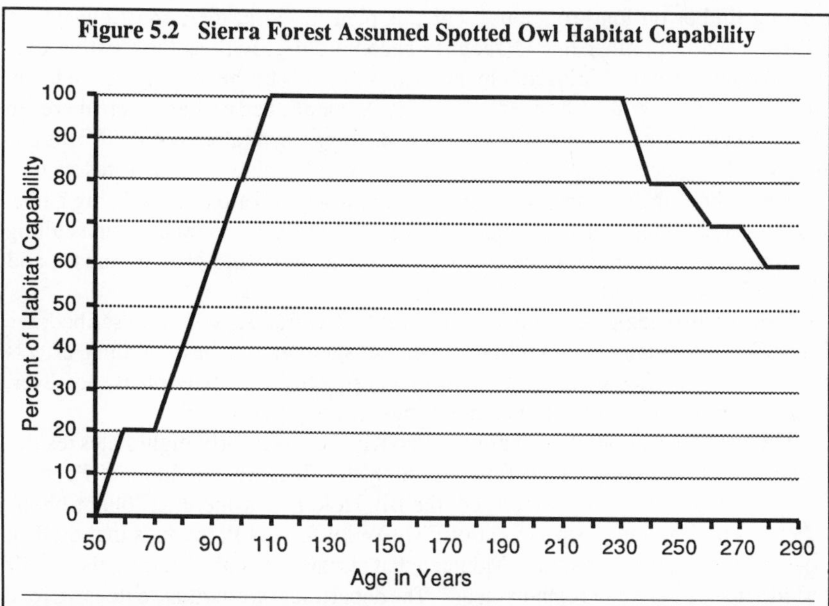

Figure 5.2 Sierra Forest Assumed Spotted Owl Habitat Capability

Although spotted owls need old-growth to breed, the yield tables used in FORPLAN often suggest that younger trees can provide sufficient habitat. This encourages the harvest of old-growth to produce more owls, when in fact such cutting will be detrimental to the owl and other old-growth dependent species.

Source: CHEC, "Review of the Sierra Forest Plan and Draft EIS" (Eugene, OR: CHEC, 1986), p. 13.

As a result, the designation of land as wilderness is projected to reduce spotted owl habitat because the trees grow older than 230 years (figure 5.2).[34] To maintain habitat, the yield tables indicate that old-growth must be cut — a claim that contradicts the biological needs of old-growth-dependent species.

Yield tables for the Sequoia Forest are similar except that optimum habitat is supposedly found in stands between 110 and 200 years.[35] Moreover, the tables are designed so that 2 acres of 80-year-old timber would provide better habitat than 1 acre of 200-year-old timber. Thus, despite the fact that the owl is supposed to be an "indicator species" for all varieties of wildlife that need old-growth forests, the yield tables project that heavy timber cutting will provide plenty of owl habitat.

Similar yield tables in the Shasta-Trinity FORPLAN model lead FORPLAN to conclude that the Forest Service can provide a sustained yield of habitat for twice the number of owls that the forest has targeted to protect.[36] Yet a 1982 memo by the Shasta-Trinity wildlife biologist indicates that, of known spotted owl territories that have not been selected for protection, "many have already been included in [timber] sales and eighty to ninety percent of all others are in the five year [timber sale] plan as are all of the quote optional sites." The memo adds that six protected territories have been lost in land exchanges and that efforts to find replacement territories "usually throws us into an existing or proposed timber sale."[37]

The Forest Service confidently plans to protect Shasta-Trinity spotted owls by growing timber on long rotations. For this plan to work, there must always be sufficient mature timber that is about to become old-growth as the existing old-growth is cut. Yet the biologist's memo warns that this timber does not exist today "and will not exist within the next fifty years. As mentioned earlier, where there are good pockets of timber there is a sale planned, going on, or already cut."[38] This on-the-ground perspective dramatically conflicts with the claims of the forest plan.

Throughout the country, wildlife concerns, whether for old-growth, big game, or diversity, are being used as a mandate to clearcut forests. In reality, such cutting will limit forest diversity, threaten species dependent on old-growth forests, and reduce big game populations in many areas.

Fish are also neglected by national forest managers. As will be described, the Forest Service usually assumes that timber management will have minimal effects on fish if it complies with "best management practices." Many plans, however, threaten the last remnants of once-numerous fish populations.

The last native population of grayling trout in the lower forty-eight states resides in the Big Hole River, which flows through the Beaverhead National Forest.[39] Besides being Butte's city watershed, the Big Hole is considered a "blue ribbon" fishery. The Montana Department of Fish, Wildlife, and Parks says that the Big Hole "is one of the few rivers in Montana that remains 'fishable' during the runoff period because the water remains clear." The department warns that "extensive road building and timber harvest proposed in the headwaters could increase sediment during the high water period and adversely affect recreational fish during the salmon fly hatch."[40]

Despite this, the Forest Service plans to dramatically increase road construction and logging in the Big Hole drainage. Although the Big Hole represents only about 25 percent of the forest, over 60 percent of the Beaverhead's timber harvest and over 70 percent of new road construction in the next ten years is expected to take place in that drainage.[41] The Forest Service's only response is that it will follow "best management practices," a legal term for water quality regulations written and enforced by the states.[42] Yet best management practices have never been defined for southwest Montana, where the Beaverhead Forest is located. Until they are, planners assume that promises to follow them will protect fish from the consequences of road construction and timber sales.

How Much Is Water Worth?

Watershed was the first nontimber resource to be mentioned in a congressional act relating to the national forests. The Organic Act of 1897 and the Weeks Act of 1911 each provided that a primary purpose of national forests was to aid in the regulation of stream flow. These laws were prompted by a belief that clearcutting of private lands had caused serious floods in many parts of the United States. Later laws, including the Federal Water Pollution Control Act, emphasized that water quality as well as quantity was to be an important consideration in national forest management.

Ironically, Forest Service claims of water benefits from timber cutting are based on research that indicates that clearcutting can produce exactly the kind of destabilizing effects on water yields that the national forests were created to prevent. Research in Colorado indicates that heavy clearcutting in a watershed will lead to higher flows during peak flow seasons and that the highest increases take place in high flood years. Based on this research, the Forest Service has proposed to heavily cut several Colorado watersheds to create the flood conditions that earlier generations considered a result of careless or immoral timber cutting.

A thirty-year study in the Fools Creek watershed, located in the Colorado front range, indicates that clearcutting increases spring flows, traditionally the peak water season. However, clearcutting had no effect on flows in the late summer, when irrigators and other water users suffer the greatest shortages. The study found that the average increase in flow following clearcutting was 23 percent. But in the ten years of the study that had the highest natural flows, the average increase was 30 percent. Supplemental water due to timber harvests were minimal during years of relative drought.[43]

The Forest Service ignored the finer points of these studies and assumed that the *average* increase in yields would be regularly used by farmers irrigating their crops. In fact, as Kenneth Turner of the California Department of Water Resources points out, farmers place the greatest value on water during droughty seasons and years.[44] Thus, the Forest Service should place an economic value only on water produced in years of low rainfall.

The value used for water by the Forest Service is even more questionable. In the 1980 RPA Program, the Forest Service estimated that an acre-foot of water produced by a national forest was worth about $5 to $10, depending on the region of the country. But the 1985 RPA escalated these values to $40 to $60.[45] The 1985 values were based on a study prepared by economists at Texas A&M University, which estimated the value of water delivered to irrigators.[46] Although all other RPA and forest planning resource values are values of the goods *in the forest* — that is, *not* delivered to the consumer — the Forest Service made no attempt to adjust the water values. Using delivered water values might be equivalent to using, for timber, the values of lumber sold to home builders or, for recreation, the total amount spent by recreationists on equipment, food, and transportation.

To convert water values to net willingness to pay, the Forest Service should have deducted the cost of storing and transporting the water to the user. Bruce Beattie, a coauthor of the Texas A&M study, notes that deducting these costs would often result in negative values, especially in locations where existing storage facilities are inadequate to hold existing runoff.[47]

The Forest Service also erred in using the consumer price index to convert the Texas A&M figures, which were in 1978 dollars, to 1982 dollars.[48] Irrigation water values have probably increased at about the same rate as agricultural prices, which have been increasing far more slowly than the consumer price index.

One of the conclusions of the Texas A&M study was that too much water is currently being used to irrigate agricultural land. It is ironic that the Forest Service is using this study to justify losing money on timber sales in order to produce more water to inefficiently irrigate more farmland, especially when so many other Department of Agriculture activities are aimed at reducing farm surpluses. In effect, the Forest Service is producing more water that can be used to subsidize irrigation for crops whose values are supported by federal price guarantees.[49] Despite all these subsidies, the Forest Service claims the added water is a benefit.

At the same time, timber management can have very detrimental effects on water quality. Road construction causes a great deal of sedimentation, and clearcuts on very steep slopes frequently lead to debris avalanches that can completely destroy fish habitat. These problems are often dismissed as unimportant by forest managers.

The Clean Water Act requires the Forest Service to comply with water quality standards established by the individual states. Under the law, states are to develop *best management practices* (BMPs), which forest landowners should use to protect water quality. For example, some states require that buffer strips be used to prevent soil from entering streams. Rather than evaluate whether its actions will seriously pollute the water, the Forest Service frequently relies on promises that it will comply with BMPs and therefore will have no significant impacts.

The environmental impact statement for the draft Klamath National Forest Plan assumed that any alternative that complied with best management practices would automatically have no detrimental effects on fish populations. Alternatives that cut between 180 million and 310 million board feet of timber per year were projected to produce identical numbers of fish because they all followed BMPs.

The state of California, however, has established specific water quality standards for sedimentation as well as BMPs. Standards define allowable levels of pollution, while BMPs describe the techniques that are expected to successfully meet the standards. A memo prepared by the Klamath's watershed specialist admitted that some alternatives would fail to meet the state sedimentation standards. This was unimportant, said the memo, because the standards were "rarely enforced." Failure to meet these standards might have serious effects on fish, but these were ignored by the Forest Service.

The Forest Service's official position is that it is not required to meet state water quality standards but only to follow the best management practices. If monitoring proves that such practices fail to protect water quality, the agency insists that it is allowed to continue such practices until the states change their definitions of best management practices.[50] This position is difficult to reconcile with claims that the agency is protecting water quality. A ruling by the U.S. Court of Appeals for the Ninth Circuit concludes that the Forest Service's view is incorrect, but the agency hopes to appeal to the Supreme Court.[51]

Many other forest plans rely on BMPs to safeguard water quality and fish habitat, even when no BMPs exist. Montana has not yet established best management practices for forest activities in many parts of the state, yet every Montana forest plan agrees to comply with these nonexistent rules. Montana does have a "nondegradation" standard, which the Forest Service often violates in the course of road construction. Although the law requires the Forest Service to obey this standard, no forest plan has ever mentioned it.

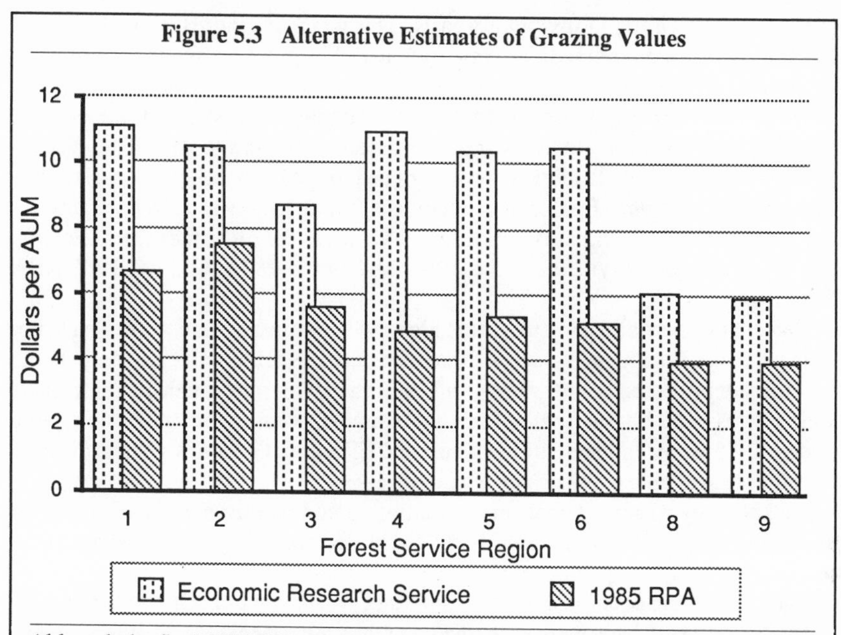

Although the final 1985 RPA Program concluded that grazing values in most regions were little more than half of the values estimated by the Economic Research Service, most forest plans are based on the ERS values.

Source: USDA Forest Service, *1985 RPA Program Final EIS* (Washington: Forest Service, 1986), p. F-7.

Below-Cost Timber Supports Below-Cost Grazing

Grazing is also used to support timber management in almost every western national forest that loses money on timber. Although the average forest spends between $2 and $6 for each animal unit month (AUM) it provides to ranchers, the Forest Service collects only $1.35 per AUM. However, forest planners admit that this is well under the fair market value for grazing and use much higher values in their computer analyses.

The 1980 RPA Program estimated that grazing values were actually in the range of $10 to $15 per AUM (figure 5.3). In 1985, however, the Forest Service and Bureau of Land Management prepared a *Grazing Fee Review and Evaluation,* which estimated that grazing on public lands is worth only about $5 to $7 per AUM.[52] The difference is critical because the average grazing costs on many national forests fall between these two ranges. If the 1980 RPA value is used, most grazing will appear to be worthwhile, but if the 1985 *Review* value is used, most will not.

Forest planners argue that studies by the USDA Economic Research Service (ERS), used in the draft 1985 RPA Program, support the higher value. ERS economists estimated the amount ranchers actually paid to private landowners for grazing rights on a county-by-county basis. This was much more thorough than the 1985 *Review*, which divided the West into just six regions and estimated grazing values for each region.

However, the *Review* allowed for the fact that public grazing lands are often of poorer quality than private lands and so cost ranchers more to use. The average amount paid to private landowners was reduced to compensate for this difference. Thus, while the *Review* figures may not be as accurate locally as ERS data, they do not consistently overestimate public grazing values as the ERS study does. For this reason, the Forest Service used the 1985 *Review* values in the final 1985 RPA Program.[53]

Where planners are quick to use the high 1985 RPA water values, they have not changed to the 1985 range values, instead using the 1980 ERS values along with the 1980 timber price trends and grazing values, both of which are much higher than the 1985 values. The Shasta-Trinity Plan, for example, is based on 1980 timber price trends, 1985 water and recreation values, and ERS grazing values — in each case, the highest available estimates of resource values.[54]

In short, by assuming that timber cutting would greatly benefit wildlife and recreation or would produce highly valued water and forage for grazing, many planners were able to persuade FORPLAN to propose high levels of timber sales despite continuing economic losses from timber management. Other planners exaggerated timber values or underestimated timber-related costs. Still others, unable or unwilling to fabricate enough data to make FORPLAN produce high amounts of timber, conjured up artificially high estimates of timber "demand" and insisted that their forests must supply their share of that demand no matter what the cost.

Clearcutting the Forest to Save It

An argument often made to support money-losing timber sales on conservation grounds is that such sales are needed to protect the forests from insects, disease, and fire. The roads built to access timber, the argument states, can later be used to give fire fighters access to wildfires. The roads also provide access for early detection and treatment of insects and disease outbreaks. Close scrutiny reveals that this argument is rarely valid.

Fire-protection costs are a function of the cost of suppression and the likelihood that a fire will take place. Given identical fires, suppression costs will be lower in an area with roads than in a roadless area because of the ease of access. However, the fires in roadless areas are very different from those in roaded areas. Most roadless area fires are caused by lightning. In roaded areas, far more fires are caused by people, either careless campers or as a result of logging operations. Human-caused fires are often larger than natural fires. In fact, most of the largest fires in this century, such as Oregon's Tillamook Burn, were caused by people.

In the Bitterroot National Forest, human-caused fires tend to be three times the size of natural fires. This may be due to the fact that fires often spread faster in developed areas than in virgin forests, which are often well insulated against fire. Or it may be due to the fact that lightning-caused fires are partially suppressed by the rain that accompanies the lightning. In either case, the cost of suppressing fires in roaded parts of the Bitterroot exceeded the costs of roadless area fire suppression even though three out of four acres of the Bitterroot are roadless.[55]

Planners in New Mexico's Gila National Forest arrived at similar conclusions, even though the climate and forest types of the Gila Forest are far different from those of the Bitterroot. Gila Forest planners estimated that total fire protection costs in developed areas would be two to eight times the costs in roadless areas. The savings is even greater in wilderness areas where the Forest Service has a "let-burn" policy.[56]

Similar conclusions were reached during the second roadless area review and evaluation (RARE II) process, where planners estimated the cost of fire protection in roadless areas before and after development. In Oregon and Washington, fire protection costs were expected to rise in almost every roadless area after the construction of roads.[57]

Insects and disease are more complicated than fire since there are so many different varieties. However, analysis frequently shows that insects and disease do little real damage to the forest and may even have beneficial effects. In other cases, insect epidemics can be serious problems, but such epidemics are actually promoted, not cured, by the Forest Service's silvicultural response.

The spruce budworm is rarely a serious problem when net timber values are negative. The budworm defoliates trees but rarely kills all the trees in a stand of timber. Instead, it successfully attacks only the weakest. It thus performs a natural thinning function, allowing the remaining trees to grow faster. Because the budworm favors low-value species such as east side Douglas-fir and true fir over high-value species such as ponderosa pine, it may actually increase the value of the forest. A recent Forest Service study found that ponderosa pine significantly benefited from budworm infestations in mixed conifer forests.[58]

The Forest Service is losing millions of dollars every year on Rocky Mountain and intermountain timber sales that are supposed to control the budworm, but such controls are probably not needed. For example, the Lincoln National Forest Plan proposes to lose $200,000 per year on its timber program, not counting the cost of roads — another $800,000 per year — and the justification for most of these losses is the reduction in spruce budworm attacks.[59] Yet a policy of spraying the forest with *Bacillis thuringiensis*, a biological agent that can control budworm epidemics, would produce similar effects and cost far less.[60]

Dwarf mistletoe is another dreaded pest that may actually be a nonproblem. Mistletoe is a plant that parasitizes the branches of many species of trees. The various species of mistletoe are very host-specific: One species attacks only ponderosa pine, while another attacks only Douglas-fir. The Forest Service argues that mistletoe reduces forest growth, and to prevent this waste, mistletoe-infested stands must be controlled through clearcutting. Below-cost timber sales designed to control mistletoe are completely pointless. Mistletoe rarely kills its host, and it does little damage to any other resource. If the timber is not worth selling, it's not worth saving from mistletoe.

Other insects and diseases are actually promoted by Forest Service timber management. *Phytophthora lateralis* is an exotic root disease that is invariably fatal to Port-Orford-cedar. The Forest Service response to the introduction of this disease into the natural range of the cedar has been to quickly log as much cedar as possible before it is infected. Yet research has shown that the disease is actually spread by road construction. *Phytophthora* spores carried from an infested site to an uninfested watershed enter the groundwater and eventually infect all Port-Orford-cedars below the road. This information has not deterred the Forest Service from its logging plans.[61]

The southern pine beetle has recently been breaking out in epidemics of record proportions. Evidence indicates that these epidemics are the direct result of monocultures — plantations of a single species — of loblolly pine, a species that is extremely susceptible to the beetle. Loblolly and other susceptible species have been favored over long-leaf pine, hardwoods, and other species that are not susceptible to the beetle, creating ideal conditions for the beetle.[62]

The Forest Service response to these outbreaks is to harvest or destroy all the trees in an infested area and plant hundreds of trees per acre of loblolly pine. Although hardwoods can provide an effective barrier to the spread of the beetle, the Forest Service in Texas has contracted to have all hardwoods in infested areas destroyed by a "tree crusher" or burned by aerial applications of a napalmlike chemical. The monocultural pine plantations that follow will create ideal conditions for future beetle epidemics.

A similar situation is developing in the northern Rockies, where mountain pine beetle epidemics can kill entire stands of lodgepole pine. Lodgepole timber sales usually lose money, but the Forest Service justifies such sales because they will supposedly preempt beetle epidemics and thus prevent the fire hazard created by large expanses of dead trees. In fact, the sales are designed to encourage establishment of new lodgepole pine forests rather than species that are less susceptible to the beetle. Thus, the Forest Service is assured of future epidemics.[63]

Recently, the Forest Service in Colorado has been aggressively promoting timber sales of aspen, which it claims are necessary to protect the scenic beauty of the state. If stands of aspen are not clearcut, says the Forest Service, they are likely to deteriorate and be replaced by spruce and other conifers. Since the aspen contributes to the fall "color" season, loss of the aspen stands is supposed to threaten the Colorado tourist industry. Therefore, the Forest Service wants to sell aspen to Louisiana-Pacific, which will make the trees into "waferboards."

What the Forest Service is neglecting to mention is that a large share of the aspen stands are actually climax forests, meaning that the aspen will replace itself as older trees die, and thus there is no danger of the aspen being replaced by conifers. Even those stands that are not climax will not be completely replaced by conifers, say Forest Service researchers, for at least 200 years. Yet the agency is proposing to cut all aspen stands in 90 years.

Ironically, the first aspen sales proposed by the Forest Service were deliberately selected stands of pure aspen, with no conifers present. These stands were most likely climax stands or, if not climax, would require several hundred years to be replaced by conifers. Thus, the alleged purpose of the aspen sales did not apply to these areas.[64]

Spruce is the only tree in Colorado national forests that comes close to paying for itself in Forest Service timber sales. Yet much of the revenue from spruce sales is dissipated on expensive reforestation designed to prevent aspen from invading the spruce clearcuts. It would be more efficient to clearcut the spruce, let aspen invade where it will, and allow aspen elsewhere to be replaced by spruce. Spruce harvests could continue — at a much lower rate — in areas where spruce had replaced aspen and matured.

Costs of Multiple-Use

Managers often claim that some national forest timber sales lose money because the costs of managing the forests are higher due to multiple-use considerations. Similar sales on private land would make money, it is claimed, and therefore the Forest Service should make the sales even if it loses money.

Proponents of this view point to a study by Forest Service economists in Montana. The economists reviewed 187 timber sales in Montana and northern Idaho between 1975 and 1981. They concluded that bids for the sales would have averaged $26 per thousand board feet more if special requirements to protect soil, wildlife, and other resources were not included in the sale contracts. The sales sold for an average of $107 per thousand board feet, so the multiple-use practices reduced sale receipts by about 20 percent.[65]

Nearly all of the costs in the study, however, are costs of mitigating the damage that timber cutting might cause nontimber resources. Thus, they should be attributed solely to timber and not to other resources. If timber receipts don't cover these and other timber-related costs, the timber sales will reduce the economic value of the forests.

Suppose that nontimber resources on a particular forest have a present net value of $100 per acre. Management of a particular acre for timber has a present net value of $50 when only timber values and costs are considered. However, such management reduces nontimber values to just $25, so the total present net value is reduced to $75. The value of the acre for nontimber resources outweighs the value for timber.

But it might be possible to mitigate the effects of timber sales on nontimber resources. Such mitigation reduces the timber values to minus $10 per acre, but the nontimber resources maintain their full value of $100 per acre. The total value of such management is thus $90 per acre — more than unmitigated timber management, but still less than nontimber management.

This conclusion changes only if the value of nontimber resources is less than the cost of the mitigation measures. In this case, the mitigation measures are inefficient. But as long as the mitigation measures are worthwhile, their cost should be charged against the value of timber. If timber-related costs, including mitigation, exceed timber returns, then the sales are economic losses.

Protecting Local Communities

The idea that the national forests exist to protect community stability is as old as the Forest Service itself. This objective has never been written into law but is entirely self-created. No congressional act relating to the Forest Service even mentioned community stability until 1944. In that year, Senator Charles McNary of Oregon convinced Congress to pass the Cooperative Sustained Yield Act, which led to the creation of the Shelton Cooperative Sustained Yield Unit in the Olympic National Forest (WA) as well as other federal sustained yield units. This law authorized, but did not require, the Forest Service to create sustained yield units to protect community stability.[66]

The 1976 National Forest Management Act also allows the Forest Service to take community stability into consideration when it decides whether to use oral or sealed bidding in timber sales. No other law allows the agency to take any action, such as selling timber at a loss, to protect community stability.[67]

Although community stability remains an important part of the Forest Service's administrative goals, so many Forest Service actions are directly contrary to protection of community stability that it cannot be the overriding reason for below-cost sales. A good example is found in the Pacific Northwest, where private timber inventories are depleted. Private second-growth timber is growing but will not be merchantable for several decades. Meanwhile, nearby national forestlands have surpluses of timber, meaning that more timber could be cut today without ever

Figure 5.4 Jobs Produced by Six National Forests

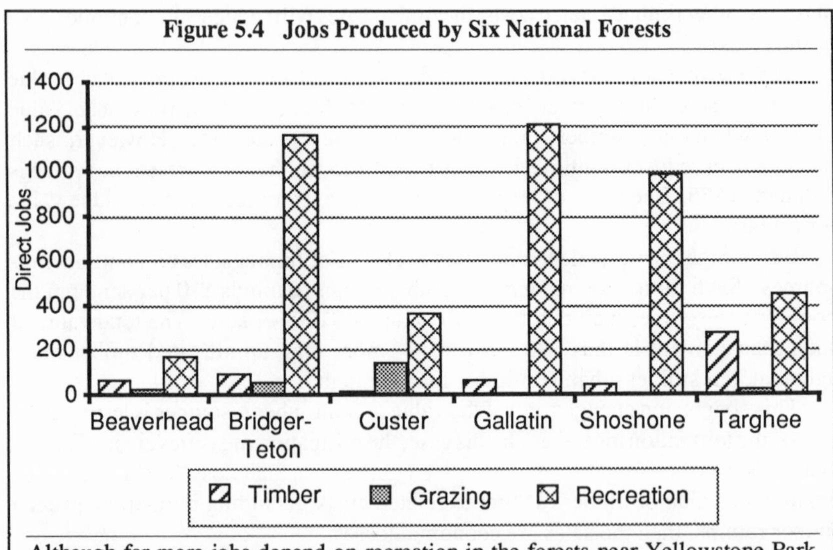

Although far more jobs depend on recreation in the forests near Yellowstone Park, forest plans and budgets for those forests reflect an emphasis on timber.

Source: CHEC, *Economic Database for the Greater Yellowstone Forests* (Eugene, OR: CHEC, 1987), p. 26.

Figure 5.5 Cost Per Job on Six National Forests

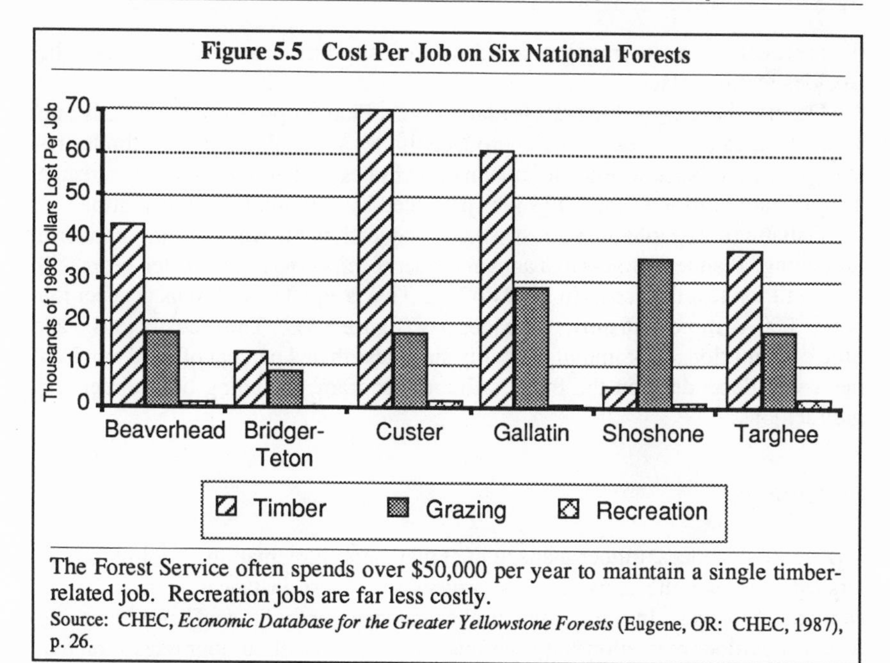

The Forest Service often spends over $50,000 per year to maintain a single timber-related job. Recreation jobs are far less costly.

Source: CHEC, *Economic Database for the Greater Yellowstone Forests* (Eugene, OR: CHEC, 1987), p. 26.

having to cut less than the long-term sustained yield capacity of the forests (see chapter 9).[68]

In these situations, community stability is not served by the Forest Service's policy of selling timber only at "nondeclining" levels. Instead, coordinating departures from nondeclining flow with the gap in private harvests would provide local communities with a steady supply of timber.[69] Despite pressures from local governments, the timber industry, and the Reagan administration, the Forest Service steadfastly refuses to consider this option.

In other parts of the West, recreation makes a greater contribution to local economies than timber. The Gallatin National Forest is one of several that border Yellowstone National Park. Due to the popularity of the park, the Gallatin and other forests in the greater Yellowstone ecosystem enjoy high levels of recreation use. In fact, forest planners estimate that Gallatin Forest recreation provides more than sixteen jobs for every job produced by cutting timber in the Gallatin.[70] Similar statistics apply to other forests bordering Yellowstone (figure 5.4).

One large source of employment is the outfitting industry, which guides and equips tourists on pack trips, river trips, and hunting expeditions. Forest Service research indicates that hunters in particular are willing to pay phenomenal amounts to hunt bighorn sheep, moose, and other wildlife that inhabit the forests adjacent to Yellowstone.[71] Outfitters rely on significant acreages of roadless land for their success. After all, few people will pay an outfitter to take them someplace that they can drive to in their car. This is one reason why outfitting is far more significant in the Rocky Mountains than on the Pacific Coast: Forest Service road-building policies decimated most roadless areas on the coast, while large roadless areas in the Rockies remain intact.

Despite these facts, the draft Gallatin Forest Plan proposes to build roads into 100,000 acres of roadless lands to produce timber and maintain the 71 timber-related jobs it estimates depend on Gallatin timber sales. [72] This proposal may threaten a significant portion of the 1,171 jobs produced by Gallatin Forest recreation.

Gallatin timber jobs are so costly that it is questionable whether they are worth protecting. Planners project that annual timber receipts over the next ten years will be $4 million less than costs in 1987 dollars. This is equal to an annual cost per job of $61,000 — more than most of the jobs pay (figure 5.5).[73] These costly jobs made little contribution to community stability since less than 2 percent of local employment was dependent on the forest's timber program, including both direct and indirect jobs.[74]

Summary

In specific instances, timber management may benefit recreation, wildlife, or watershed or reduce the costs of forest protection. These instances seem to be exceptions, however. In most cases, timber management is in conflict with scarce forms of outdoor recreation and beneficial only to forms of outdoor recreation that are already available in abundance.

Limited timber sales may benefit some species of big game, but intensive timber management reduces the amount of cover needed by game for shelter from predators and temperature extremes. Timber management also tends to fragment forests and reduce habitat quality for rare or endangered species.

Timber management can increase water flows, but such increases take place mainly during high-flow seasons and years of high flows. Forest Service estimates of the value of such water increases appear to be greatly exaggerated.

Timber cutting can address certain forest protection problems, but most often timber management spreads or perpetuates, rather than prevents, insects and disease. Even where timber cutting can reduce insect problems, other options, such as prescribed burning, can have the same effect at a lower cost. Roads needed for timber cutting also increase the likelihood that public access will increase fire management costs.

In short, money-losing timber sales cannot be justified by the benefits those sales provide for other resources. There must be some other explanation for below-cost sales.

Notes

1. "The Role of 'Below Cost' Timber Sales in National Forest Management," unpublished paper on file at the Forest Service, Washington, DC, 16 August 1984, p. 1.
2. Ibid.
3. USDA Forest Service, *National Forest Roads for All Uses* (Milwaukee, WI: Forest Service, 1985), p. 2.
4. USDA Forest Service, *Chequamegon National Forest Plan Draft EIS* (Park Falls, WI:

Forest Service, 1985), p. C-1.
5. Ibid., p. B-22.
6. CHEC, "Review of the Chequamegon Forest Plan and Draft EIS" (Eugene, OR: CHEC, 1985), pp. 1–3.
7. USDA Forest Service, *Bitterroot National Forest Plan Draft Environmental Impact Statement* (Hamilton, MT: Forest Service, 1985), p. II-5.
8. Ibid., p. II-7.
9. John V. Krutilla, "Statement Before the Subcommittee on Forests, Family Farms, and Energy of the House Agriculture Committee," June 1985, p. 9.
10. "Planning Memo No. 1840-A," memo on file at the Hoosier Forest supervisor's office, Bedford, IN, p. 1.
11. USDA Forest Service, *Review Draft of Analysis of the Management Situation*, unpublished document on file at Forest Service office in Bedford, IN, November 1982, p. 99 (handwritten note).
12. Jack Ward Thomas, ed., *Wildlife Habitats in Managed Forests—The Blue Mountains of Oregon and Washington* (Portland, OR: Forest Service, 1979), p. 118.
13. USDA Forest Service, *Lincoln National Forest Plan Final Environmental Impact Statement* (Alamagordo, NM: Forest Service, 1985), p. 231.
14. Thomas, *Wildlife Habitats*, p. 114.
15. CHEC, "Review of the Draft Carson Forest Plan and EIS" (Eugene, OR: CHEC, 1985), p. 15.
16. CHEC, "Review of the Draft Beaverhead Forest Plan and EIS" (Eugene, OR: CHEC, 1985), p. 20.
17. USDA Forest Service, *Gallatin National Forest Proposed Forest Plan* (Bozeman, MT: Forest Service, 1985), p. III-44.
18. Interview with James Devitt, Gallatin planning team leader, May 1985.
19. Keith J. Hammer, "An On-Site Study of the Effectiveness of the U.S. Forest Service Road Closure Program in Management Situation One Grizzly Bear Habitat, Swan Lake Ranger District, Flathead National Forest, Montana," report prepared for Swan View Coalition and Resources, Ltd., November 1986, p. 1.
20. "Procedure for Estimating the Effects of Alternative Forest Plan Prescriptions on Five Flathead National Forest Grizzly Bear Productivity Areas," unpublished memo on file at the Flathead National Forest, Kalispell, MT.
21. Eric D. Forsman, *A Preliminary Investigation of the Spotted Owl in Oregon*, masters thesis on file at Oregon State University, Corvallis, OR, 127 pp.
22. Jerry F. Franklin et al., *Ecological Characteristics of Old-Growth Douglas-Fir Forests* (Portland, OR: Forest Service, 1981), p. 5.
23. E. V. Bull, "Ecology of the Pileated Woodpecker in Northeast Oregon," *Journal of Wildlife Management* 51:472–481.
24. Thomas A. Hanley, *Relationships Between Sitka Black-Tailed Deer and Their Habitat* (Portland, OR: Forest Service, 1984), p. 14.
25. John Schoen, Olof Wallmo, and Matthew Kirchhoff, "Wildlife Needs Virgin Forests," *Forest Planning* 2(4):7–9, 18.
26. "Diversity Technical Report," unpublished paper on file at Coconino National Forest, Flagstaff, AZ.
27. Jerry F. Franklin et al., "Modifying Douglas-Fir Management Regimes for Nontimber Objectives," *in* Chadwick Dearing Oliver, Donald P. Hanley, Jay A. Johnson, eds., *Douglas-Fir Stand Management for the Future: Proceedings of a Symposium* (Seattle, WA: University of Washington, 1986), pp. 373–379.
28. Thomas, *Wildlife Habitats*, pp. 122–123.
29. Barbara Anderson, "Owl Wars," *Forest Watch,* 6(8):12–14.
30. Larry D. Harris, *The Fragmented Forest* (Chicago, IL: University of Chicago Press, 1984), 211 pp.

31. Franklin et al., *Old-Growth Douglas-Fir*, p. 40.
32. Comment made by Olaf Wallmo at Old-Growth Symposium, Eugene, OR, February 1982.
33. CHEC, "Review of the Four Northern Arizona Forest Plans" (Eugene, OR: CHEC, 1986), pp. 36–39.
34. CHEC, "Review of the Sierra Forest Plan and Draft EIS" (Eugene, OR: CHEC, 1986), p. 13.
35. CHEC, "Review of the Draft Sequoia Forest Plan and EIS" (Eugene, OR: CHEC, 1986), p. 19.
36. USDA Forest Service, *Shasta-Trinity National Forests Plan Draft Environmental Impact Statement* (Redding, CA: Forest Service, 1986), p. II-130.
37. Kenneth Coop, "SOMA Narrative Comments," unpublished memo on file at the Shasta-Trinity Forests supervisor's office, Redding, CA, 14 July 1982, p. 2.
38. Ibid., p. 6.
39. "Fisheries," draft comments on Beaverhead supplemental DEIS prepared by Montana Department of Fish, Wildlife, and Parks, p. 3.
40. Ibid., p. 6.
41. CHEC, "Review of the Draft Beaverhead Plan," p. 13.
42. USDA Forest Service, *Beaverhead Forest Plan Final Environmental Impact Statement* (Dillon, MT: Forest Service, 1986), p. IV-79.
43. C. A. Troendle and R. M. King, "The Effect of Timber Harvest on the Fools Creek Watershed, 30 Years Later," *Water Resources Research* 21:1915–1922.
44. Kenneth M. Turner, "Water Loss and Potential Water Salvage in Southern California," presentation at American Geophysical Union, San Francisco, CA, December 1984, p. 6.
45. USDA Forest Service, *1985-2030 Resources Planning Act Program Final Environmental Impact Statement* (Washington, DC: Forest Service, 1986), p. F-5.
46. Michael Frank and Bruce Beattie, *The Economic Value of Irrigation Water in the Western United States* (College Station, TX: Texas A&M University, 1979), p. 87.
47. Interview with Bruce Beattie, April 1984.
48. USDA Forest Service, *1985-2030 Resources Planning Act Program Final Environmental Impact Statement*, p. F-5.
49. Richard Domingue, "Timber and Water," *Forest Watch* 7(1):19–21.
50. Letter from Peter Myers to Lee Thomas on file at USDA, Washington, DC, 29 August 1986, 4 pp.
51. *Northwest Indian Cemetary Ass'n* v. *Peterson*, 565 F. Supp. 586, 590 (N.D. Cal. 1983).
52. USDA Forest Service and USDI Bureau of Land Management, *Grazing Fee Review and Evaluation* (Washington, DC: Forest Service and BLM, 1986), p. 15.
53. USDA Forest Service, *1985-2030 Resources Planning Act Program Final Environmental Impact Statement*, p. F-7.
54. CHEC, "Review of the Shasta-Trinity Forests Plan and Draft EIS" (Eugene, OR: CHEC, 1986), p. 22.
55. CHEC, "Review of the Bitterroot National Forest Draft Plan and EIS" (Eugene, OR: CHEC, 1985), p. 15.
56. CHEC, "Review of the Draft Gila Forest Plan and EIS" (Eugene, OR: CHEC, 1985), p. 11.
57. Randal O'Toole, *An Economic View of RARE II* (Eugene, OR: CHEC, 1978), pp. 50–53.
58. Clinton E. Carlson et al., *Release of a Thinned Budworm-Infested Douglas-Fir/Ponderosa Pine Stand* (Ogden, UT: Forest Service, 1985), p. 7.
59. CHEC, "Review of the Draft Lincoln Forest Plan and EIS" (Eugene, OR: CHEC, 1985), pp. 16–17.
60. Ibid., pp. 11–12.

61. Thomas Lawson, *Management of Port-Orford-Cedar and Its Influence on* Phytophthora *Root Rot* (Eugene, OR: CHEC, 1983), pp. 5–8.
62. Thomas Lawson, "Pine Monoculture Leads to Pine Beetles," *Forest Watch* 7(2):15–18.
63. Thomas Lawson, "Mountain Pine Beetle: Forest Pest or Forest Regulator" *Forest Watch* 7(5):19–23.
64. CHEC, "Review of the Grand Mesa-Uncompahgre-Gunnison Forests Plan" (Eugene, OR: CHEC, 1985), pp. 9–10.
65. Robert E. Benson and Michael J. Niccolucci, *Costs of Managing Nontimber Resources When Harvesting Timber in the Northern Rockies* (Ogden, UT: Forest Service, 1985), p. 19.
66. Cooperative Forest Management Act of 1944 (16 USC 568c, 568d).
67. National Forest Management Act of 1976 (16 USC 472a).
68. John H. Beuter, K. Norman Johnson, and H. Lynn Scheurman, *Timber for Oregon's Tomorrow* (Corvallis, OR: Oregon State University, 1976), p. 19.
69. Oregon State Board of Forestry, *Forestry Program for Oregon* (Salem, OR: Oregon Department of Forestry, 1977), p. 37.
70. USDA Forest Service, *Gallatin National Forest Plan Draft EIS* (Bozeman, MT: Forest Service, 1985), p. B-46.
71. Dennis M. Donnelly and Louis J. Nelson, *Net Economic Value of Deer Hunting in Idaho* (Ft. Collins, CO: Forest Service, 1986), p. 1; Cindy F. Sorg and Louis J. Nelson, *Net Economic Value of Elk Hunting in Idaho* (Ft. Collins, CO: Forest Service, 1986), p. 1; Dennis M. Donnelly et al., *Net Economic Value of Recreational Steelhead Fishing in Idaho* (Ft. Collins, CO: Forest Service, 1985), p. 1; John B. Loomis et al., *Net Economic Value of Hunting Unique Species in Idaho: Bighorn Sheep, Mountain Goat, Moose, and Antelope* (Ft. Collins, CO: Forest Service, 1985), p. 1.
72. USDA Forest Service, *Gallatin Draft EIS*, p. II-103.
73. CHEC, "Review of the Gallatin National Forest Plan and EIS" (Eugene, OR: CHEC, 1985), p. 21.
74. USDA Forest Service, *Gallatin Draft EIS*, pp. III-7, B-46.

Part II

Forming the Diagnosis

Chapter Six
Why Does the Forest Service Lose Money?

In 1983, some environmentalists were talking with foresters who work for the Washington State Department of Natural Resources, and the conversation turned to large, old-growth trees. "Have you seen our old-growth western redcedar on the Olympic Peninsula?" one of the foresters asked. "It's the largest in the world — 18 feet in diameter and 198 feet tall!" The tree, it was explained, was in a timber sale but had been overlooked by DNR samplers estimating the sale volume. When Rayonier, the purchaser, found the tree, the company realized it was of record size and graciously agreed not to cut it.

"There was another tree nearby that was almost as big," said the foresters, "about 17 feet in diameter and 195 feet tall. But they cut that one down." The environmentalists were aghast at the thought that all the trees surrounding the record cedar had been logged, leaving none for protection from wind. Wasn't the department afraid the record tree would blow down? "Oh, no," said an official. "Cedar resists decay. If it blows over, we'll be able to salvage it all." The fact that one of the largest trees in the world might be lost was unimportant because the wood, at least, would be used.

This attitude is a classic example of *timber primacy*, the notion that timber is the most important forest product and must be produced at any cost. Nonforesters who frequently work with the Forest Service are familiar with this viewpoint: It can be seen in the Forest Service brochure that says, "We harvest timber because it is needed for man's survival." It can also be seen in the bumper sticker that reads "Wood is good."

Many environmental groups, believing that timber primacy is the reason why the Forest Service makes so many below-cost and environment-damaging timber sales, call for prescriptive legislation to correct these problems. But is timber primacy really the major motivation for the Forest Service's apparently irrational behavior? The answer to this question is key to developing successful reforms because reforms that fail to address Forest Service motives are likely to be subverted by the bureaucracy. A number of economists suggest that *budget maximization* is the actual goal of any bureaucracy. If this is true, prescriptive legislation might have little effect on Forest Service behavior.

The Timber Religion

Evidence of timber primacy dates to the nineteenth century in the Bureau of Forestry, the agency that preceded the Forest Service. Bernard Fernow, the first director of the bureau, argued, "Our civilization is built on wood. From the cradle to the coffin, in some shape or other, it surrounds us."[1] The Society of American Foresters, which was started by Forest Service founder Gifford Pinchot, was more succinct: "No wood, no agriculture, no manufacture, no commerce."[2] Richard Alston points out that timber primacy may have seemed valid in 1900, when "most things were made of wood and most Americans depended on wood for fuel."[3] But per capita wood consumption in the United States has declined significantly since that era, and total U.S. wood consumption has remained steady and, in fact, has never exceeded the 1908 level.[4]

Nevertheless, the belief in timber primacy seems to endure. David Clary, the former chief historian of the Forest Service, says that the Forest Service "is different from other bureaucracies, which operate on a body of policy that continually adjust to the changing world. The Forest Service, however, has something more like a religion. From the beginning it has perceived itself as fulfilling a sacred mission to provide wood to the world in order to avert the evils of a 'timber famine.' "[5]

William Duerr, who has been called the "father of forest economics," describes timber primacy as the belief that "timber is the chief product of the forest; all else that comes from the forest is a by-product, of secondary interest: water, forage, wildlife, and the rest, including recreation. Indeed, people are a nuisance in the forest. Wood is, and will always be, a necessity, for it has no true substitutes. Its consumption is assured; its consumers may be taken for granted. In fact, there is going to be a shortage of timber, and the central problem in forest management is the biological and engineering problem of growing more timber."[6]

It is tempting to believe that timber primacy is responsible for below-cost timber sales and other inefficient Forest Service policies. Yet timber primacy is only one of many beliefs held by turn-of-the-century foresters. Most of the others have disappeared over the years. Those who believe that timber primacy rules the Forest Service need to explain why these other beliefs have disappeared.

For example, Pinchot and other early foresters strongly opposed large-scale clearcutting. As David Clary notes, "Clearcutting was uncommon on the national forests until after World War II, except in the case of Douglas-fir. As foresters came to exert increasing influence over logging in this century, selective harvesting was favored as a corrective to the destructive effects of old industrial logging." In the 1930s, the Forest Service even applied selection cutting to Douglas-fir forests and other forests that today are often claimed to "require" clearcutting. [7] Some foresters still believe that selection cutting is far superior to clearcutting,[8] yet the Forest Service has adopted clearcutting as the primary method in almost every national forest.

Another strong belief was that clearcutting caused floods. To convince Congress to pass a law permitting the Forest Service to purchase lands and create national forests in the East, Gifford Pinchot proclaimed that floods in the Ohio

Valley were "due fundamentally to the cutting away of the forests on the watersheds of the Allegheny and Monongahela Rivers."[9] Today, most Forest Service officials reject the idea that clearcutting causes floods. Ironically, research that indicates clearcutting does increase water flows is used to justify more clearcutting in arid parts of the country.

A third major belief was that fires must be kept out of forests at any cost. At one point, a Forest Service critic noted that the Forest Service made "complete fire protection, in all circumstances and regardless of conditions, not a theory but a religion, an *idee fixe*."[10] The Smokey Bear "prevent forest fires" campaign is a vestige of this belief. But today, after many years of resistance by Forest Service officials, prescribed burning is used in many regions of the country, the policy of suppressing all wildfires by ten o'clock of the morning they are detected is no longer mandatory, and many wilderness wildfires are allowed to burn freely.

Finally, the Forest Service long fought for the authority to regulate private timber management, claiming that such regulation was "essential to the public interest."[11] But in 1953, the Forest Service "decided to drop the recurrent crusade for regulation, which gained nothing but the antagonism of a powerful part of the Forest Service constituency."[12]

While timber primacy was always important to some Forest Service officials, it may not have always played the major role that it does today. Timber remained important to those who were in charge of timber, but many forest officials had other priorities. During the 1930s, Forest Service chiefs William B. Greeley and Ferdinand Silcox actively supported expansion of the wilderness system.[13] In 1945, Clary reports, the Forest Service's director of timber management complained that "timber resource management in [Region 5] has been regarded as a side issue to improvements, protection, and recreation. Apparently actual forestry practice in R-5 can be handled by low paid subordinates."[14] In 1948, his successor reviewed Region 2 activities and "was most displeased to learn that forest supervisors and district rangers spent most of their time working with those who held grazing permits, leaving timber work to inadequately trained or supervised junior staff. He reported that 'the Region is preoccupied with range and watershed issues and programs.' The fact that the region was in the early stages of a 'range war' over grazing regulation and stock reduction and that grazing was its outstanding headache apparently were irrelevant."[15]

By 1966, the Washington office was more oriented toward timber, but some regions still were not. Clary notes that "in the Southwest Region [Arizona and New Mexico], the Washington office was quick to complain: 'One cannot help but feel that behind a facade of multiple use many people in the Region continue to be primarily concerned with grazing and recreation, often at the expense of other resources and uses.' Washington wanted timber to be emphasized," Clary concludes, but the region did not agree.[16] Apparently, timber primacy became particularly dominant within the Forest Service only during the 1950s and 1960s.

Beliefs in selection cutting, in the need to suppress wildfires, that clearcutting causes floods, and in the need to regulate private lands were once held by agency officials with religious fervor. Yet today, these ideas have all but disappeared.

Timber primacy, which was not universal in the Forest Service's early years, now appears dominant. Perhaps timber primacy has grown because it supports other Forest Service goals — such as a larger budget — while the other beliefs have diminished because they did not.

Laws of Economics

The budget-maximization hypothesis is an outgrowth of the fundamental economic assumption that people's decisions are strongly influenced by the incentives affecting the decision makers. Change the incentives, and the decisions change. Change the people but not the incentives, and in the long run the decisions remain pretty much the same.

Using this assumption, economists have built models of "consumer behavior" and "producer behavior" which assume that consumers and producers are influenced by prices. All goods and services are created from some combination of basic factors of production — raw materials, labor, and capital. According to the *scarcity theory of value*, the value of goods and services is determined by the scarcity of the factors of production required to form them relative to the demand for those factors.

This scarcity theory of value distinguishes mainstream economics from Marxian economics. Marx, along with other early economists, believed that goods were valuable because of the amount of labor required to make them. A diamond, for example, was thought to be more valuable than water because diamonds required more labor to mine and process than water. Thus, Marxian economics is based on a *labor theory of value* and overlooks the fact that diamonds are more valuable than water simply because there is so much water and there are so few diamonds.

Many resource economists and managers also implicitly or explicitly rely on a *resource theory of value*. Gifford Pinchot once said, "There are just two things on this material earth — people and natural resources," thus disregarding the role played by capital in making goods and services available to consumers. People who advocate that energy be used to measure the value of ecosystems and the cost of manufacturing processes are ignoring the fact that nonenergy resources, not to mention labor and capital, also have a role in determining values.

Raw materials, labor, and capital are general terms for what are, in fact, a diverse set of items. Black walnut and lodgepole pine are both raw materials, but the first is far more valuable than the second both because it is scarcer and because people are willing to pay more for the beautiful dark wood. Obtaining capital — that is, a loan — in 1981 was more difficult than in 1979 because the Federal Reserve Bank had tightened up the money supply and raised interest rates. People who wanted a loan in 1981 had to be willing to pay more interest than in 1979. And the labor of a highly skilled artist is more valuable than that of an unskilled worker if many people are willing to pay for the artist's works.

Willingness to pay, then, is the measure of how consumers value various goods and services. The theory of consumer behavior assumes that most consumers have an income and make practical decisions every day about how they spend that income and what they are willing to pay for various goods. The most ardent wilderness advocate still uses paper and wood products, while many timber industry executives use the wilderness for hunting, hiking, and horseback riding — indicating that each is willing to pay for both wood and wilderness.

Some people are willing to pay more for various goods than others, and most people will buy a larger quantity of a good if the price is lower. Thus, the total amount of a good that will be purchased increases if its cost declines. A set of points representing the price required to convince consumers to buy various quantities of goods is called a *demand curve.* When graphed with price on the vertical axis and quantity on the horizontal axis, such a curve slopes downward and to the right (figure 6.1).

The theory of producer behavior assumes that producers want to maximize their profits, and thus they will be sensitive to prices they can obtain for their goods and the costs of producing them. Producers must pay the costs of raw materials, capital, and labor to manufacture goods and services. These costs may increase as the volume of goods produced increases. For example, to harvest more timber may require entering roadless areas or cutting trees on steeper slopes, which are more costly to remove. Thus, producers will increase production only if the price people are willing to pay is greater than the cost of production. A set of points representing the price required to convince producers to manufacture various quantities of goods is called a *supply curve.* When graphed with price on the vertical axis and quantity on the horizontal axis, a supply curve slopes up.

Figure 6.1 Supply and Demand Curves

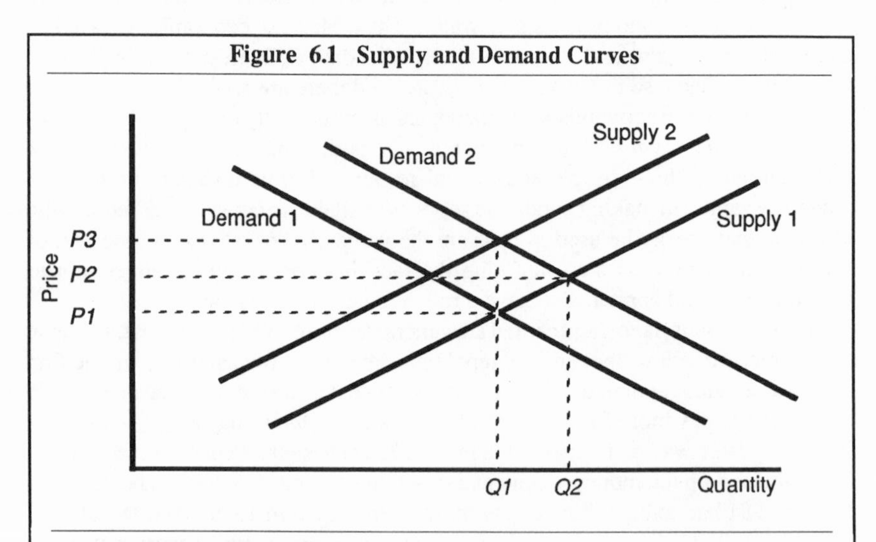

If demand increases from demand 1 to demand 2, perhaps because of a change in tastes or an increase in population, then prices will increase from *P1* to *P2* and the quantities consumed will change from *Q1* to *Q2*. If at the same time supplies decrease from supply 1 to supply 2, perhaps because some timberlands are transferred to farm, urban, or other nontimber use, then prices will increase to *P3* but quantities will remain at about *Q1*.

The intersection of the supply and demand curves defines the price and quantity at which, given the existing distribution of income, everyone is satisfied. At a higher quantity, producers are paying more for production than they are receiving for the goods and are failing to sell the goods because consumers won't pay the high prices. At a lower price, producers are unwilling to supply all that consumers are willing to buy.

Although supply and demand curves are relatively straightforward, demand in particular is frequently confused with the *quantity demanded.* For example, the Forest Service often issues statements like, "The U.S. demand for timber in the year 2030 will be 70 billion board feet." This statement is meaningless in the context of figure 6.1. Does the Forest Service mean that demand will be 70 billion board feet no matter what the price of timber is? This would be absurd, as high prices would obviously reduce the quantity demanded. Does the Forest Service mean that the quantity demanded will be 70 billion board feet at current prices? If so, it is not too important, as prices are sure to increase if the quantity demanded is greater than the quantity supplied. Or does the Forest Service mean that the intersection of the supply and demand curves will result in consumption of 70 billion board feet? If so, there is no problem since producers will obviously have the incentive to produce 70 billion board feet.

A *perfect* market—where there are so many producers and consumers that none has an influence on the price or quantity of the good being produced, where consumers have perfect knowledge about the relative quality of the various goods, and where the costs of producing a good are entirely absorbed by the producers, who also receive all the benefits from selling the good — is self-regulating. If a change in technology leads to a reduction in the cost of production, new producers will increase production so that more goods are available. Production will also increase if a change in consumer tastes leads to an increase in the value of a particular good.

A perfect market is considered *efficient* because raw materials, capital, and labor will tend to be put to their highest and best uses, and the total net value of all the goods produced will be as high as possible. Given the existing distribution of income, it is also considered *equitable* because people are paying for the goods they want and not for goods that other people want.

Such a perfect world does not exist, of course. The main reason why the national forests are publicly owned is that people like Gifford Pinchot and Theodore Roosevelt were able to convince Congress that government ownership would overcome the problems of imperfect markets. Since that time, a large body of economic literature has documented the failures of the marketplace to efficiently and equitably allocate goods. These problems will be discussed in chapter 13.

Economic theories of consumer and producer behavior can be tested by examining markets for various goods and services. However, it is immediately clear that such tests do not apply to the Forest Service. Although the national forests produce a wide range of goods and services, a fundamental assumption of the theory of producer behavior does not apply to the Forest Service: that the producer is interested in maximizing profits. The Forest Service clearly does not maximize its profits and has repeatedly stated that profit maximization is not its goal.[17]

Budget Maximization

What is the goal of a public agency such as the Forest Service? William Niskanen, a former member of the President's Council of Economic Advisors, examined public agencies in his book *Bureaucracy and Representative Government*. He notes, "Among the several variables that may enter the bureaucrat's utility function are the following: salary, perquisites of the office, public reputation, power, patronage, output of the bureau, ease of making changes, and ease of managing the bureau." All but the last two items, says Niskanen, are a positive function "of the total *budget* of the bureau during the bureaucrat's tenure in office" (emphasis in original).[18] Moreover, while it may be easier to manage and change a smaller bureau than a larger one, at any given time the bureaucrat will perceive that management and change can be made easier with a larger budget. For these reasons, Niskanen suggests that agencies attempt to maximize their budgets.

Other organizational experts agree. According to Peter Drucker, a specialist in the management of large organizations, public agencies are "based on a 'budget' rather than being paid out of results. . . . And 'success' in the public-service institution is defined by getting a larger budget rather than attaining results. Any attempt to slough-off activities, therefore, diminishes the public service institution. It causes it to lose stature and prestige. Failure cannot be acknowledged. Worse still, the fact that an objective has been attained cannot be admitted."[19]

For top managers, larger budgets mean greater prestige. For middle managers, larger budgets mean more people on their staff, and this generally provides them with higher salaries. For lower managers, larger budgets mean greater opportunities for advancement.

A good example of the budgetary influence on the Forest Service can be seen in the Resources Planning Act (RPA). RPA was passed for the specific purpose of providing stable budgets for improved national forest management. Ten years later, when people examined RPA to see if it was successful, they didn't ask, "Did RPA improve national forest management?" Instead, they asked, "Did RPA increase Forest Service budgets?"[20] Increased budgets is the measure of success.

Budget maximization may also be a motivation for managers of private institutions, but for private entities in the free marketplace, budget maximization and profit maximization are nearly identical. Where they differ, pressures from stockholders and other owners will tend to make managers conscious of the profit-maximization goal. For government agencies, budget maximization is much different from either profit maximization or maximization of net social benefits.

Ronald Johnson, an economist at Montana State University, recently reviewed the Forest Service bureaucracy and concluded that budget maximization explains its actions better than other explanations of its activities. Capture by clients, adherence to a conservation ethic, and economic efficiency each fail to explain certain Forest Service policies or activities as well as budget maximization.[21]

Johnson points out that below-cost timber sales allow the agency to "spread harvesting activities across political jurisdictions." Reviewing timber sale levels in the twelve western states, Johnson concludes that "the volume sold annually tends to be proportional to the inventory of sawtimber on each national forest" rather than to its value. With the support of interest groups such as the timber industry, the Forest Service can successfully convince members of Congress to vote for budget increases because they have a political incentive to do so. Johnson adds that Congress encourages this process with *logrolling*, the practice of allowing senators and representatives who are most concerned about timber or other pork barrel to dominate the subcommittees that appropriate the funds.

Johnson also considers the Multiple-Use Sustained-Yield Act to be a strategy aimed at increasing budgets, noting that funding for recreation, fish and wildlife, and soils and water have been increased by a factor of ten since passage of the act. In addition, he says, "Multiple use management requires considerably more planning and service-related activities than would be needed if areas were designated on the basis of dominant use. Since 1964 the annual harvest on the national forests has not increased, but appropriations for timber sales administration have more than doubled in real terms. That increase is largely attributable to increased monitoring and planning necessitated by muliple-use management."

Below-cost timber sales, grazing, and recreation have another purpose: Artificial underpricing of the resources increases the quantity demanded for those resources (figure 6.2). Shortages appear to exist, and all user groups have a common interest in lobbying Congress for increased budgets to alleviate those shortages.

Others agree that the Forest Service is motivated by its budget. Peter Emerson and Robert Turnage, economists with the Wilderness Society, note that "managers of public agencies, such as the Forest Service, have an incentive to maximize annual

Figure 6.2 Shortages Due to Underpricing of a Resource

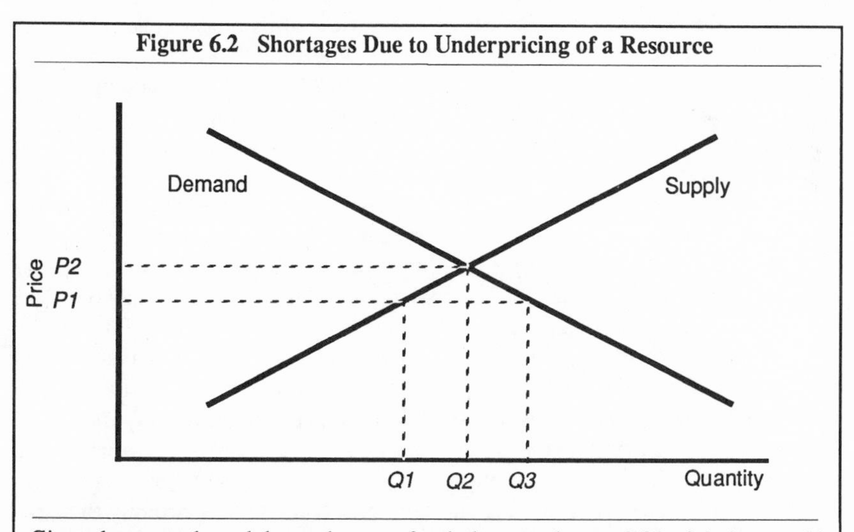

Given these supply and demand curves for timber, a volume of *Q2* of timber would normally sell for *P2*. But if the Forest Service sells a significant share of its timber below-cost, prices will fall to *P1*. This will increase the quantity demanded to *Q3* while other timber owners will reduce their outputs so that the quantity supplied falls to *Q1*. The result is an artificial shortage of timber.

budgets and staff positions under their control." With the increased demand for timber after World War II, they say, "the Forest Service discovered a reliable political prescription for increased budgets and staffing. Congress readily appropriated funds for the Forest Service to hire a large work force of foresters and road engineers to prepare timber sales and design the necessary road system."[22]

Although Clary argues that timber primacy is the main motive behind Forest Service behavior, his work provides support for budget maximization as well. He notes that "The Service had developed a major program to sell multiple use (and a larger budget for itself) in 1960, calling it 'A Servicewide Plan to Gear Multiple Use Management of the National Forests to the Nation's Mounting Needs.' "[23] Clary also describes the RPA planning process as "an answer to a bureaucrat's prayer. It superficially realized longstanding visions in the Forest Service of a comprehensive national forestry plan based on a comprehensive inventory, with the Service producing both according to its own judgment. More practically, RPA authorized the endless generation of paperwork, which necessitated the expansion of the agency's work force."[24]

Clary's claim that the Forest Service is "unlike other bureaucracies" is contradicted by his statement that "talk of balanced programs, interdisciplinary planning, and multiple use could not alter the Forest Service's focus on timber. It was, after all, a federal bureaucracy, and accordingly it was affected by the flow of power in the government. By its nature as a revenue-producing activity, timber management figured in national economic calculations. Congress and successive presidents reinforced the Service's timber orientation by calling for expanded national-forest sales to support economic growth or provide federal revenues."[25]

In fact, the Forest Service's near-religious view of its mission would be considered predictable by Peter Drucker. Drucker points out that public agencies exist "to 'do good.' " "This means," says Drucker, that the agencies "tend to see their mission as a moral absolute rather than as economic and subject to a cost/benefit calculus. Economics always seeks a different allocation of the same resources to obtain a higher yield. . . . In the public service institution, there is no such thing as a higher yield. If one is 'doing good' then there is no 'better.' Indeed, failure to attain objectives in the quest for a 'good' only means efforts need to be redoubled."[26]

Advocates of the budget-maximization hypothesis do not believe that bureaucrats are looting the Treasury for personal gain. In fact, an agency may be a budget maximizer even though not everyone in the agency is particularly interested in — or conscious of interest in — maximizing budgets. "Consider the probable consequences for a subordinate manager who proves without question that the same output could be produced at, say, one-half the present expenditure," says Niskanen. "In a profit seeking firm, this manager would probably receive a bonus, a promotion, and an opportunity to seek another such economy; if such rewards are not forthcoming in a specific firm, this manager usually has the opportunity to market his skills in another firm.

"In a bureau," continues Niskanen, "at best, this manager might receive a citation and a savings bond, a lateral transfer, the enmity of his former colleagues, and the suspicion of his new colleagues. Those bureaucrats who doubt this probability and have good private employment alternatives should test it . . . once."[27]

Some agencies, for a time, may try to maximize something other than their budgets. If they maintain this behavior, however, they are likely to be replaced or eclipsed by other agencies that prove superior at convincing Congress to allocate funds to them.

A Process of Natural Selection

The assertion that the Forest Service is motivated by a desire to maximize its budget does not imply Forest Service officials are unscrupulous individuals who are trying to steal from the taxpayers. Most, if not all, Forest Service employees are honest, well-intentioned people who truly believe that what they are doing is right. Most foresters and other professionals within the agency strongly believe in their version of a conservation ethic and base many of their decisions on this ethic. Yet the agency as a whole can still be strongly motivated by its budget.

A bureaucracy such as the Forest Service can probably be best understood by considering it a giant but primitive multicelled organism, perhaps something like algae. When placed by itself in a nutrient culture, the organism grows as long as there is food available. But few organisms are so lucky to live by themselves in a sea of unlimited food. Instead, they typically must compete with other organisms.

To compete for food, organisms evolve to take the best advantage of their environment. Competition with other organisms spurs this evolution, and individual cells that are unable to utilize the available food will die and be replaced by others that are better able to do so. No conscious intelligence is necessary for natural selection. At any given time, the organism doesn't know which change will make it best able to compete. Many different changes may take place, but only a few will be successful.

The Forest Service and other federal agencies are similar to such organisms, and the food they compete for is the federal budget. The Forest Service has been fairly successful at capturing a steady and growing share of that budget. To do so, it has evolved with the times and with the environment created for it by congressional acts. For example, when it was advantageous to claim that clearcutting causes floods because such claims inspired Congress to fund the creation or purchase of more national forests to prevent such floods, the Forest Service made such claims. These claims were actually contrary to evidence gathered by Forest Service researchers, evidence that was suppressed by Forest Service administrators.

Years later, the claim that clearcutting causes floods was disadvantageous to the Forest Service because clearcutting led to increased Forest Service budgets. Thus, the Forest Service proclaimed that clearcutting did not cause floods. Even later, when the Forest Service was fighting people who claimed that it was wasting tax dollars by selling timber at a loss, clearcutting was suddenly found to increase water supplies — not enough to cause floods but enough to help irrigate the arid West. Of course, the fact that most of the water produced by clearcuts came at peak water seasons, and that even more was produced during peak water years, did not mean that clearcutting causes floods.

This evolutionary view of organizations is supported by Peters and Waterman's book, *In Search of Excellence.* They conclude that "We believe that the truly adaptive organization evolves in a very Darwinian way. The company is trying lots of things, experimenting, making the right sorts of mistakes; that is to say, it is fostering its own mutations. The adaptive corporation has learned quickly to kill off the dumb mutations and invest heavily in the ones that work."[28] Where the private companies discussed by Peters and Waterman evolve to increase their profits, a public service organization like the Forest Service evolves to increase its budget.

Making Careers in Timber

It is possible, then, that the Forest Service's conservation ethic — including timber primacy — has been shaped by the bureaucracy's tendency to maximize its budget. In 1945, the Forest Service and its conservation ethic were very different from what they are today. There was little demand for national forest timber, and the demand that existed was met almost exclusively with selection cutting and uneven-aged management. Large chunks of the National Forest System had been set aside by Forest Service officials as primitive areas or wilderness.

In the next quarter century, however, Forest Service policies and its conservation ethic changed dramatically. This resulted largely from increasing demand for national forest timber. The agency as a whole took advantage of this demand to dramatically increase its budget — well beyond the point of economic efficiency or maximum social good. At the same time, selection cutting was almost completely replaced by clearcutting, and primitive areas were declassified so that access roads could be built and the timber sold.

Historian Clary hints that the timber dominance so visible after 1950 had its source in budgets and the career goals of Forest Service employees. He notes that "timber was such an active program by 1952 that any ambitious young forester could see that in the Forest Service, timber was where careers were to be made." He adds that timber was important for two reasons: "Philosophically, national forest timber was the future salvation of the nation, to become available when other supplies ran out. Pragmatically, sales receipts justified the existence of the forests and earned notice for foresters who produced them."[29]

The agency's conservation ethic evolved to rationalize the new timber policies. The change was not sudden but rather occurred as a process of natural selection: Forest managers who believed in activities that tend to increase the agency's budget were promoted over those who did not. Suppose, for example, that two district rangers have differing views on ideal forest management. One believes that wilderness is an appropriate use of the forest and wants to preserve a large portion of his district for recreation, wildlife, and natural beauty. The other believes that timber management is needed to maintain a healthy forest for a wide variety of wildlife and that timber-related roads are valuable to recreationists and other forest users.

In the 1930s, when there was little demand for national forest timber, the first ranger might have appeared credible to agency leaders. Congress would at least appropriate funds to protect the wilderness, including money for recreation and wildlife protection. By 1950, however, the timber-oriented ranger would make a much more significant contribution to the agency's budget. Demand for national forest timber was growing, and timber sales would increase the agency budget by encouraging appropriations from approving members of Congress.

The timber ranger's supervisor need not have any particular beliefs about the superiority of timber over wilderness. But the timber ranger, having a bigger budget, would be able to hire a larger staff. This would also, in effect, increase the number of staff members under the ranger's supervisor. The overall budget and prestige of the timber branch of the agency would be enhanced. The timber ranger and his or her supervisor would be seen as "go-getters" and would be favored for promotion over the wilderness ranger and his or her supervisor.

Over time, the top ranks of the Forest Service would come to be dominated by people who sincerely believe that timber management is good — even vital — for recreation, wildlife, watershed, community stability, and "the survival of mankind." Day-to-day decisions are made by people who share a conservation ethic — but these people are in the position to make the decisions simply because their conservation ethic coincides with maximizing the agency's budget.

Multiple Clientism

A third popular explanation for Forest Service below-cost sales is that the agency has been "captured" by the timber industry and simply does what the industry wants. Conflicts between the industry and the Forest Service over such issues as nondeclining flow and the minimum harvest age of timber are too numerous to make this likely.

Paul Culhane, a political economist, has examined this theory in detail and concludes that it should probably be "discarded altogether" because the interest groups influencing the Forest Service are too numerous for any one to dominate.[30] However, he substitutes a new theory of his own, which he calls "multiple clientism." "The influence of the agencies' balanced constituencies is consistent with national agency policies," he notes. By being "responsive" to local constituents, forest administrators are "conforming to informal agency and professional norms of responsiveness (presumably a desirable trait for public officials in a democracy)."[31]

This does not conflict at all with the budget-maximization hypothesis. As Culhane notes, "Clientele support is one of the four basic sources of bureaucratic power, along with officers' professional expertise, cohesion *esprit de corps*, and mastery of the game of bureaucratic politics." The Forest Service and BLM "use rational and sophisticated tactics to manipulate their political environments," including "play[ing] their more extreme constituents off against each other to reinforce the agencies' preferred middle course."[32] Although Culhane appears to believe that the agencies' goals are "multiple use" (which, in the case of the Forest Service, may seem equivalent to timber primacy), it is clear that these same tactics could be used to maximize agency budgets.

The debate over whether timber primacy or budget maximization is the true explanation of Forest Service behavior is almost a question of the chicken or the egg. Does the Forest Service believe in timber primacy only because it maximizes its budget? Or does the Forest Service just happen to increase its budget because of its belief in timber primacy? Fortunately, it is possible to examine Forest Service behavior to determine which explanation is more realistic.

Chapter 7 will examine the Forest Service timber sales process and predict how the agency will behave if it is a budget maximizer or a timber primacist. Chapter 8 will demonstrate that Forest Service actions indicate that it is a budget maximizer. Chapters 9 and 10 will show that Forest Service policies on other issues, such as nondeclining flow, clearcutting, and wilderness, also suggest that it is a budget maximizer.

Notes

1. Bernard Fernow, *Economics of Forestry* ((New York, NY: Thomas Crowell, 1902), pp. 21–22.
2. Committee for the Application of Forestry, "Forest Devastation: A National Danger and a Plan to Meet It," *Journal of Forestry* 17(12):915.
3. Richard M. Alston, *The Individual and the Public Interest* (Boulder, CO: Westview Press, 1983), p. 19.
4. David Clary, *Timber and the Forest Service* (Lawrence, KS: University Press of Kansas, 1986), p. 41.
5. Ibid., p. xi.
6. William A. Duerr and Jean B. Duerr, "The Role of Faith in Forest Resource Management," *in Man and the Ecosystem* (Burlington, VT: City Printers, 1971), p. 51.
7. Clary, *Timber*, p. 181.
8. Peter A. Twight and Leon S. Minckler, *Ecological Forestry for the Northern Hardwood Forest*, (Washington, DC: National Parks and Conservation Association, 1972), 12 pp; Gordon Robinson, "Excellent Forestry," *Conifer*, winter 1969.
9. Ashley L. Schiff, *Fire and Water* (Cambridge, MA: Harvard University Press, 1962), p. 120.
10. Jenks Cameron, *The Development of Government Forest Control in the United States* (Baltimore, MD: Johns Hopkins Press, 1928), p. 320.
11. Clary, *Timber*, p. 147.
12. Ibid., p. 149.
13. James M. Glover, *A Wilderness Original: The Life of Bob Marshall* (Seattle, WA: The Mountaineers Books, 1986), p. 265.
14. Clary, *Timber*, p. 120.
15. Ibid., p. 123.
16. Ibid., p. 173.
17. "The Role of 'Below Cost' Timber Sales in National Forest Management," unpublished paper on file at the Forest Service, Washington, DC, 16 August 1984, p. 1.
18. William Niskanen, *Bureaucracy and Representative Government* (Chicago, IL: Aldine and Atherton, 1971), p. 38.
19. Peter Drucker, *Innovation and Entrepreneurship* (New York, NY: Harper and Row, 1985), p. 179.
20. Dennis C. LeMaster, "Forest Service Funding Under RPA," *Journal of Forestry* 80(3):161–3.
21. Ronald N. Johnson, "U.S. Forest Service Policy and Its Budget," *in Forestlands: Public and Private* (San Francisco, CA: Pacific Institute for Public Policy Research,

1985), pp. 103–133.
22. Peter M. Emerson and Robert W. Turnage, "The Great Tongass Timber Heist: Vested Interests and Obsolete Policies on the National Forests," paper on file at the Wilderness Society, Washington, DC, p. 14.
23. Clary, *Timber*, p. 184.
24. Ibid., p. 189.
25. Ibid., p. 193.
26. Peter Drucker, *Innovation*, p. 179–180.
27. William Niskanen, *Bureaucracy*, p. 39.
28. Thomas J. Peters and Robert H. Waterman, Jr., *In Search of Excellence* (New York, NY: Harper and Row, 1982), p. 115.
29. Clary, *Timber*, p. 151.
30. Paul J. Culhane, *Public Lands Politics — Interest Group Influence on the Forest Service and the Bureau of Land Management* (Baltimore, MD: Johns Hopkins University Press, 1981), p. 332.
31. Ibid., p. 335.
32. Ibid., p. 336.

Chapter Seven
The Timber Sale Process

In December 1986, I visited a Forest Service regional office to collect data on national forest expenditures and receipts. The Forest Service keeps these data in two different offices: Expenses are in the budget office and receipts are in the fiscal office. After copying the budgetary data, I visited the fiscal office to get receipts for fiscal year 1986, which ended in September. I mentioned to one of the accountants that I had just come from the budget office as well.

"You know," he said, "some people get our budgets and receipts and try to compare the two. I don't understand why they do that. I used to work for a private company, where it was important that income be greater than expenses, and there we took pains to see that they did so. But the Forest Service isn't in the private sector. We're a government agency, and we don't have to make a profit."

It may not be true that government agencies "don't have to make a profit," but it is certainly true that the national forest timber sale process gives forest managers little incentive to compare benefits and costs. If the Forest Service is a budget maximizer, as suggested by Ronald Johnson, then several congressional acts actually give the agency an incentive to lose money on sales.

Among the laws regulating Forest Service timber sales are the Organic Act of 1897, which allows the Forest Service to sell timber for not less than appraised value.[1] The Deposits from Brush Disposal Act of 1916 allows the Forest Service to collect deposits for the disposal of brush following timber sales.[2] The Knutson-Vandenberg Act of 1930 authorizes the Forest Service to keep a share of timber sale receipts to spend on reforestation.[3] The National Forest Roads and Trails Act of 1964 gives the Forest Service the right to require timber purchasers to build roads to the timber sales, the costs of which can be credited against the purchase price of the timber.[4]

Most recently, section 14(h) of the National Forest Management Act creates a revolving fund for timber salvage sales; the receipts from salvage sales can be used to prepare and administer new salvage sales and to engineer and design purchaser-built roads to the sales.[5] NFMA also repealed parts of the 1897 Organic Act, but the appraised value language was retained by NFMA.[6] The use of Knutson-Vandenberg funds was also expanded by NFMA to include precommercial thinning, wildlife habitat improvement, and other activities.[7]

These well-intentioned laws were based on the idea that timber is an important use of the national forests. To produce timber, the Forest Service must fund roads, brush disposal, and reforestation. To prevent dead and dying trees from going to waste, the Forest Service must fund timber salvage sales. All of these laws work together to provide funding for these activities in a relatively noncontroversial manner. Few people would argue against reforestation. Even fewer would argue against a funding mechanism designed to ensure that the Forest Service has the money to reforest land after the timber has been harvested.

When the Organic Act was repealed in 1976, Congress retained the appraised value requirement and noted that this was intended to ensure "that the United States will obtain fair market value for timber and forest products."[8] The Forest Service had always interpreted "appraised value" to mean "fair market value," and, in fact, Forest Service regulations note that "the objective of national forest timber sale appraisal is to estimate fair maket value."[9] The *Forest Service Manual* defines fair market value as "the price acceptable to a willing buyer and seller, both with knowledge of the relevant facts and not under compulsion to deal."[10] Since no reasonable seller would perpetually agree to sell goods at a loss, fair market value appears to preclude below-cost timber sales.

However, the manual directs timber sale officers to use an appraisal method called the *residual pricing system*. This system allows the sale of timber below cost despite the fair market value requirements. Residual pricing begins with published prices that mills are paying for logs and subtracts the costs of cutting and transporting trees to the mill.[11] The remainder becomes the *indicated advertised rate*.

Sometimes the indicated rate is negative — the costs of logging and transportation are more than the trees are worth. To avoid selling trees for negative amounts, the Forest Service also has a *base rate*, which is the lowest amount for which timber may be sold. At minimum, the base rate is equal to $0.50 per thousand board feet. The *advertised rate*, or minimum bid price, is the greater of the indicated advertised rate and the base rate. If the advertised rate is greater than the indicated advertised rate, the sale is called a *deficit sale*, not because the Forest Service expects to lose money but because the timber purchaser is expected to lose money.

Timber managers also estimate the cost of roads that purchasers will be required to build. The road cost, sometimes called *purchaser credits*, may be credited against the high bid price of the sale but not against the base rates. The purchaser is required to build the road even if purchaser road costs are greater than the high bid price minus the base rate — another example of a deficit sale.

Problems with Appraisals

The Forest Service timber sale pricing policy suffers from several problems. First, nowhere in the appraisal process does the Forest Service consider the costs to the taxpayers of placing timber for sale. Actual sale preparation and administration costs average between $10 and $15 per thousand board feet. Purchaser-built roads must be engineered and designed, usually at a cost averaging $10 to $20 per thousand board feet. The Forest Service often uses appropriated funds to build roads when the road costs are so high that purchasers will not buy timber if they are required to build the roads. These costs add up to far more than the $0.50 per thousand board feet base rate. Yet the Treasury collects no more than the base rate from many sales.

Second, the federal government loses when purchasers bid deficit sales up to the amount of purchaser credits plus base rates. The Forest Service collects the same amount of cash whether the timber sells for base rates or for base rates plus purchaser credits or for any price in between. But at the higher prices the Treasury must turn more money over to counties. This is because the counties in which the national forests are located receive 25 percent of timber sale receipts, *including* purchaser credits. Counties also receive 25 percent of the dollars the Forest Service retains under the Knutson-Vandenberg Act but no portion of the money collected under the Brush Disposal Act.

Suppose that the minimum price of a sale is $4,000, out of which the Forest Service keeps $3,000 for reforestation, but the sale contract requires the purchaser to build roads costing $24,000. If the purchaser bids $4,000, the counties would get $1,000 and the Forest Service $3,000. But if the purchaser bids $24,000, counties get $6,000 and the Forest Service $3,000. The sale costs the purchaser the same and the Treasury collects no more than $1,000 in either case, since the rest of the sale is paid for in roads rather than in cash. Thus, the Treasury must pay counties $5,000 more for the higher bid sale out of income that doesn't exist. A 1984 report by the staff of the House Appropriations Committee called such bids *wooden dollar* bids.[12]

The Forest Service's complicated appraisal process was once termed an "exercise in futility" by the staff of the House Appropriations Committee.[13] After reviewing a number of recent timber sales, the staff concluded that "appraisals do not reflect fair market value."[14] This implies that the Forest Service is not behaving like a prudent seller when it offers to sell timber for less than the cost of accessing the timber and arranging timber sales. But does this mean that the Forest Service is acting irrationally? Perhaps, from its own viewpoint, below-cost sales are very rational. A model of a hypothetical national forest can be built to find out if this is true.

Modeling a Forest

The model of the hypothetical national forest will assume that funds for sale preparation, administration, and road engineering are appropriated by Congress. Purchasers may be required to build roads or roads can be built with road construction funds. Some of the purchaser payments for timber may be retained by the Forest Service as brush disposal deposits or K-V deposits for reforestation and forest improvements. For simplification, timber salvage sales are left out, but this omission will not change the results.

The Hypothetical National Forest has three major timber types, arbitrarily called ponderosa pine, Douglas-fir, and lodgepole pine. Table 7.1 displays some of the characteristics of each forest type. Ten million board feet of each type can be harvested on a nondeclining basis at a cost of $35 per thousand board feet, including sale preparation and roads but not brush disposal, reforestation, or forest improvements. Half the road costs of $20 per thousand board feet can be paid for with purchaser credits, but the other half is for engineering and design and must be

Table 7.1 Hypothetical Forest Timber Values and Costs

Forest Type	*Ponderosa Pine*	*Douglas-Fir*	*Lodgepole Pine*
Annual Sale Volume MBF	10,000	10,000	10,000
Value Per MBF	$75	$25	-$25
Sale Cost Per MBF	35	35	35
Net Return	40	-10	-60
Total Annual Return	$400,000	-$100,000	-$600,000

In this hypothetical example, ponderosa pine forests represent above cost timber, Douglas-fir forests are below cost, and lodgepole pine forests are negatively valued.

appropriated by Congress.

Stumpage values for each forest type are estimated using the residual pricing formula. The ponderosa is worth an average of $75 per thousand board feet to purchasers, the Douglas-fir only $25, and the lodgepole minus $25. The ponderosa pine can thus be classified as above-cost timber, since it is worth more than the costs to the Forest Service of accessing and selling it. The Douglas-fir is below-cost timber. Given appropriations from Congress, below-cost timber can be sold, but the returns are exceeded by the appropriations. The lodgepole pine is negatively valued, which is even worse than below cost. Table 7.2 breaks down the costs of timber management to the Forest Service. All values and costs are in dollars per thousand board feet, and while they are typical of timber values and costs in parts of the West, they are all hypothetical.

Given this information, several ways of managing the Hypothetical Forest can be imagined based on differing forest objectives. The forest might be managed to maximize budgets or to maximize timber sales. The forest might also be managed to maximize profits for the U.S. Treasury or to maximize returns to counties. Counties, of course, collect 25 percent of timber receipts, including purchaser credits and Knutson-Vandenberg deposits but not including brush disposal deposits. The results of this model are shown in table 7.3 and figure 7.1.

Maximizing Profits

If the Hypothetical Forest is managed to maximize profits, no below-cost or negatively valued timber will be sold. Instead, sales will be limited to 10 million board feet per year of ponderosa pine. Depending on yields of second-growth timber and problems with natural reforestation, artificial reforestation might be preferred. If so, the Forest Service would have a budget of $15 per thousand board feet for sale preparation and $10 per thousand board feet for road engineering and design. Managers would also keep $5 per thousand for brush disposal, $5 per thousand for reforestation, and $10 per thousand for forest improvements, for a total budget of $45 per thousand, or $450,000 plus $100,000 in purchaser credits.

Counties get 25 percent of all receipts except brush disposal. Since ponderosa is worth $75 per thousand board feet and brush disposal costs $5 per thousand,

Table 7.2 Hypothetical Forest Timber Costs

Activity	*Cost Per MBF*
Natural Reforestation	$1
Artificial Reforestation	5
Forest Improvements	10
Brush Disposal	5
Sale Preparation & Administration*	15
Road Engineering*	10
Road Construction*	10

Costs are in dollars per thousand board feet. Starred (*) items are included in sale costs in table 7.1. The remaining items are investments in future management and so are not counted as sale costs.

counties get $17.50 per thousand board feet or $175,000. The U.S. Treasury collects all timber receipts other than purchaser credits and what the Forest Service keeps, or $45 per thousand board feet. Deducting appropriations and payments to counties leaves $2.50 per thousand board feet, or $25,000 for the Treasury.

Maximizing County Payments

If the Forest were maximizing returns to counties, Douglas-fir would be brought under management. Although the Hypothetical Forest Douglas-fir is below-cost timber, counties receive a percentage of gross (except for brush disposal) rather than net receipts. With the addition of Douglas-fir, appropriations to the Forest Service total $500,000, while the forest's share of timber receipts is $400,000 plus $200,000 in purchaser credits. Total receipts are $1 million, but brush disposal deposits are $100,000, so the counties' share is 25 percent of $900,000, or $225,000. After the Forest Service, purchaser road credit, and counties' shares of receipts are deducted, the Treasury has $375,000 in cash, out of which it must pay the $500,000 appropriation. Thus, the Treasury loses $325,000.

Table 7.3 Costs and Returns from Alternative Management Plans

	Treasury Returns	*County Payments*	*Agency Budget*	*Timber Sales*
Timber Receipts	$750	$1,000	$550	$250
Retained by Forest	200	400	520	235
Appropriations	250	500	893	1,050
Total Forest Budget	550	1,100	1,430	1,285
Payments to Counties	175	225	105	25
Return to Treasury	450	400	13	15
Net to Treasury	$25	-$325	-$985	-$1,060

All values are in thousands of dollars per year. Each alternative represents the maximization of a different objective—returns to the Treasury, county payments, budgets, and timber outputs.

Figure 7.1 Forest Service, County, and U.S. Treasury Returns

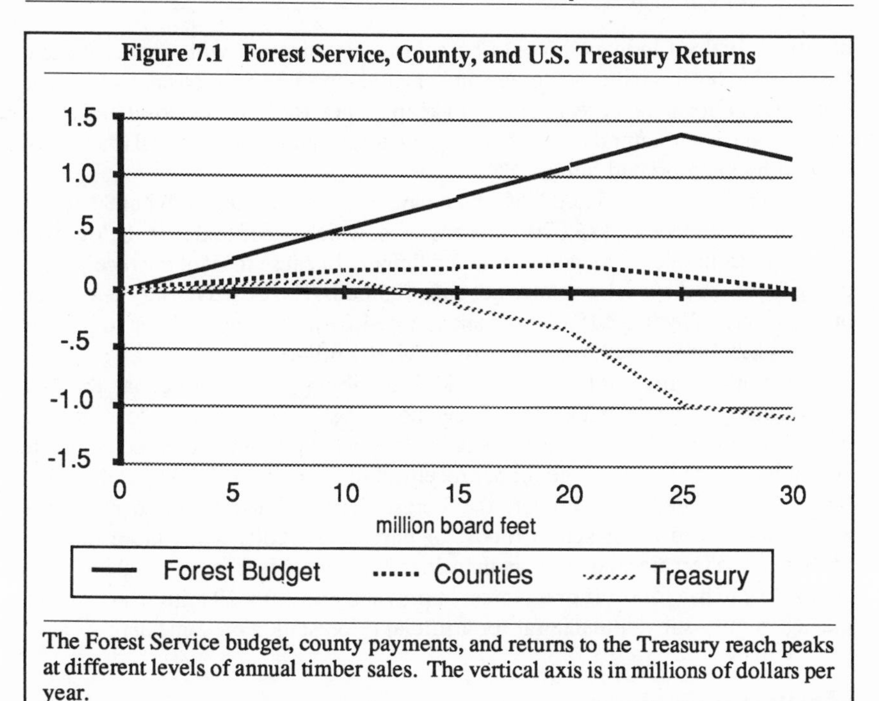

The Forest Service budget, county payments, and returns to the Treasury reach peaks at different levels of annual timber sales. The vertical axis is in millions of dollars per year.

Note that all roads are still built with purchaser credits. At $25 per thousand board feet, Douglas-fir is not valuable enough to pay for the brush disposal, reforestation, and forest improvements (which total $20 per thousand board feet) plus purchaser credits ($10 per thousand board feet). The Forest Service could reduce its reforestation or forest improvement receipts or it could seek congressional appropriations for road construction. Instead, it is assumed that the Hypothetical Forest combines the more valuable ponderosa pine in the same sales as Douglas-fir. This may be the path of least resistance since it requires neither a budgetary sacrifice nor congressional action.

From an accounting viewpoint, purchaser credits have the same effect on the Treasury as appropriated funds. From a political viewpoint, the Forest Service may consider purchaser credits cheaper than appropriated roads because it can argue to Congress that the timber, rather than tax dollars, is paying for the roads. From a budgetary viewpoint, the Forest Service is probably indifferent between purchaser credits and appropriated road funds since it has to hire about the same number of engineers to oversee construction in either case.

Maximizing Timber

If the Hypothetical Forest were run by timber primacists, its managers would try to sell all of the lodgepole pine as well as the ponderosa and Douglas-fir. The negatively valued lodgepole could be sold in combination with ponderosa and Douglas-fir, but this reduces average timber values to $8.33 per thousand board feet. This return allows the Forest Service to keep money for brush disposal and some artificial and some natural reforestation but no forest improvements. All roads must be built by appropriated funds.

Total receipts are $250,000, of which the Forest Service keeps all but $0.50 per thousand board feet, or $15,000. The counties, which benefitted from below-cost sales, lose from sales of negatively valued timber: Because most of the receipts are going for brush disposal, counties get only $25,000. The U.S. Treasury is an even bigger loser, collecting $15,000 but spending $1.05 million and having to pay the counties $25,000 as well, for a net loss of $1.06 million.

The timber primacist budget totals $1.285 million. This is more than the $1.1 million budget when only Douglas-fir and ponderosa are sold, but it is less than the maximum possible budget. As sales of the lodgepole pine are increased from 0 to 10 million board feet, average timber receipts fall from $50 to $8.33 per thousand board feet. To sell the timber, the Forest Service must begin to substitute appropriated road construction funds for purchaser credits when more than 23 million board feet are sold. If more than 26 million board feet are sold, sale receipts are too low to pay for some or all forest improvements. After 29 million board feet, managers must substitute natural for artificial reforestation on some acres.

Maximizing Budgets

As shown in figure 7.1, the Hypothetical Forest budget is maximized at 26 million board feet of sales. Managers retain $520,000 in timber receipts and obtain another $893,000 in appropriated funds. Purchaser credits are $17,000, so the total forest budget is $1.43 million, or $145,000 more than the timber primacist budget.

This model makes it possible to predict Forest Service behavior given any of these four objectives. If the Forest Service is a profit maximizer, it will not sell any below-cost or negatively valued timber. If the Forest Service is attempting to maximize returns to counties, it will sell below-cost timber but not negatively valued timber.

Sales of negatively valued timber would suggest that the Forest Service is either a budget maximizer or a timber primacist. As a budget maximizer, however, it would limit sales of negatively valued timber to protect its own budget, ensuring that timber receipts paid for brush disposal, reforestation, and forest improvement activities.

Budget-maximizing managers might also attempt to maximize reforestation costs, using artificial reforestation whenever justified by timber receipts. In some cases, they might be willing to forego receipts for forest improvements or use less expensive reforestation methods if justified by increases in appropriations for timber sales and roads. They would certainly use purchaser credits for most road construction where timber values were high enough, but they would apply appropriated funds in preference to reducing their own receipts where timber values were low.

This model has shown that the Forest Service can maximize its budget using several strategies: selling below-cost and negatively valued timber, increasing the costs paid by K-V funds, brush disposal (BD), and timber salvage sales (TSS), and adjusting purchaser road credits (PC). Each of these strategies deserves in-depth discussion.

Cross-Subsidizing Timber

Most national forests contain a wide variety of timber types and classes. Timber on steep slopes costs purchasers more to log than timber on gentle slopes, so they will pay less for it. Certain species of timber are more valuable than others, so purchasers will pay more. Marginal revenues would be obtained from low-valued timber on steep slopes.

Almost all national forests have some timber whose value is negative — the cost to the purchaser of logging the timber is greater than its value to the purchaser. Most forests also have some timber whose value is positive. The Forest Service can limit itself to selling only positively valued timber, but the total volume of timber sold can be increased by combining high-valued timber with low-valued timber in the same sales. In turn, this strategy augments the Forest Service budget by increasing the amount of funds for K-V activities and brush disposal which the agency collects.

Consider a district ranger who is contemplating two potential timber sales. Each sale has a million board feet of timber, but one is composed mainly of ponderosa pine — a very valuable species — while the other is composed mainly of lodgepole pine — a low-value species. As a result, the ponderosa unit is appraised at $100 per thousand board feet, while the lodgepole pine unit has an indicated advertised rate of negative $80 per thousand board feet (table 7.4). The district silviculturist estimates that reforestation will cost $9,500 for each sale. Sale preparation and

Table 7.4 Hypothetical Cross-Subsidization

	Unit 1	*Unit 2*	*Both Units*
Volume (MBF)	1,000	1,000	2,000
Appraised Value Per MBF	$100.00	-$80.00	$10.00
K-V Deposit Per MBF	9.50	9.50	9.50
Base Rates Per MBF	10.00	10.00	10.00
Total Price	$100,000	$0	$20,000
Retained by Forest	9,500	0	19,000
Payments to Counties	2,500	0	5,000
Net Return to Treasury	$65,500	$0	–$4,000

Of two hypothetical units of timber, one is worth $100 and the other minus $80 per thousand board feet. If unit 1 is sold alone, receipts total to $100,000 of which the Forest Service retains $9,500, counties receive $25,000, and the U.S. receives the rest. Unit 2 cannot be sold alone because of its low value. If both units are sold, receipts total to only $20,000, of which counties receive $5,000. The Forest Service doubles its income, but the U.S. loses $69,500 on unit 2.

administration costs will be $15 per thousand board feet, and no new roads are required. How will a rational ranger design the sales?

If the ponderosa pine unit is sold, the Forest Service collects $100,000 dollars, out of which the Forest Service keeps the $9,500 K-V reforestation deposit. The remaining $90,500 is sent to the U.S. Treasury, which pays $25,000, or 25 percent of the timber sale receipts, to the local counties. Sale preparation and administration costs $15,000, so U.S. taxpayers net $50,500 on the sale.

No one is likely to buy the lodgepole pine at base rates because the timber has such a low value. Although the appraiser estimates the value of the wood on the stump is negative $80 per thousand board feet, the minimum base rates are $0.50 per thousand board feet. Thus, the Forest Service estimates that the purchaser will lose at least $80,500, and even if the appraisal underestimates the timber value, it is unlikely that any purchaser could profit from this sale. Thus, neither the counties nor the Treasury nor the Forest Service collect any money.

However, instead of selling the two units separately, the ranger can combine them into one sale. The minimum bid price for the lodgepole can be increased by $90 per thousand, allowing the Forest Service to keep $9.50 per thousand for the K-V fund and still give the fifty cents per thousand base rates to the Treasury. The bid price for the ponderosa can be reduced by $90 to make up this difference. The entire sale can then be sold for $10 per thousand board feet, or a total of $20,000 for the 2 million board feet.

Counties receive $5,000, or $20,000 less than if the ponderosa is sold alone. The Treasury collects only $0.50 per thousand, or $1,000, but must find $5,000 to pay the counties, so it actually loses $4,000. The Treasury also pays $30,000 for sale preparation and administration. Federal taxpayers net a negative $34,000 and lose the $65,500 they could have made if the ponderosa pine were sold alone. The Forest Service doubles its income, collecting $19,000 for reforestation activities.

Note that the sale could still work if the lodgepole pine were worth as little as minus $99 per thousand board feet. The average bid price would then be fifty cents per thousand, or the minimum base rate, which means that the Forest Service would collect no money for reforestation or other K-V activities. A timber primacist would not hesitate to arrange such a sale, but a budget maximizer would insist on protecting the budget for management. A key feature that distinguishes a budget maximizer from a timber primacist would be a mechanism to ensure that sales of negatively valued timber do not jeopardize the managers' budgets.

If all sales in a forest include both positively and negatively valued timber, all sales will sell for the same average value even though some timber is actually much more valuable. Stands of high-valued timber would commonly *cross-subsidize* stands of low-valued or negatively valued timber, even though the stands might be miles apart. In this case, the economic rule of equating marginal revenue — which is the revenue from the least valuable timber — with marginal cost — which is the cost of selling the least valuable timber — becomes irrelevent to the forest manager.

Instead, the budget-maximizing manager attempts to equate average revenues with the marginal costs of K-V activities, brush disposal, timber salvage sales, and purchaser credit roads (figure 7.2). The result is a much higher level of harvests than is economically efficient, including the sale of much below-cost and negatively

Figure 7.2 Budget-Maximizing Behavior

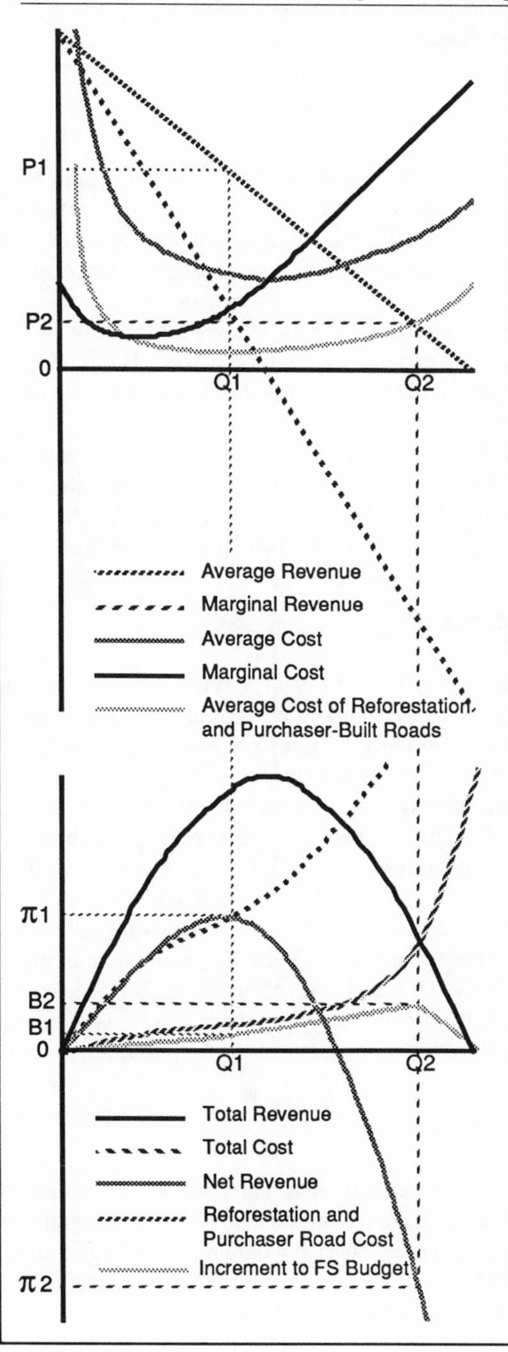

Increasing timber sales requires entry into less valuable and more costly timber, so average revenues decline while average costs increase. Private businesses try to equate marginal revenues with marginal costs — quantity *Q1* and price *P1*. This maximizes profits at point *π1*.

Budget maximization suggests that national forest managers are interested in increasing the share of revenues they retain from sales. Cross-subsidization makes all sales return the same average amount, allowing managers to equate average rather than marginal revenues and costs.

Budgets can thus be increased by increasing sales to *Q2*, where average revenues equal the average cost of reforestation plus purchaser-built roads. Beyond this point, falling timber prices would reduce the agency's budget. Prices at *Q2* are *P2*, reducing profits to a very negative *π2*. Yet the Forest Service adds *B2* to its budget, rather than just *B1*.

Budgets can be further increased by reducing purchaser-built road costs, perhaps by convincing Congress to appropriate road construction funds. This reduces the average cost of purchaser roads, moving *Q2* to the right.

valued timber. In contrast, the timber primacist would allow average revenues to fall as low as $0.50 per thousand board feet (or even less if possible), leading to an even higher sale level but a lower budget.

Maximizing Costs

Even with cross-subsidization, the average value of timber in some forests might be much greater than the K-V, brush disposal, and timber salvage sale costs. To increase its budget, the Forest Service would use management techniques that increase these costs. For example, reforestation is one of the most important costs of forest management. In most national forests, the K-V reforestation budget is larger than the combination of BD, TSS, and other K-V expenditures together. A profit-maximizing entity would attempt to keep reforestation costs low. But a budget-maximizing agency would use expensive reforestation techniques — up to the point where marginal reforestation costs equal average timber receipts.

Based on this principle, a budget-maximizing Forest Service would vary reforestation costs according to the value of the timber being sold. High-value timber or timber sold when wood prices were high would be charged greater K-V deposits than low-value timber or timber sold when wood prices were low.

Appropriated Road Construction

Purchaser credits represent an easy way for the Forest Service to gain access to roadless areas. Purchasers can be required to build roads, while the Forest Service is saved the trouble of gaining appropriations from Congress other than those needed to cover the costs of road engineering and design.

In national forests with high timber values, reforestation, brush disposal, and similar costs may not be high enough to equal average revenues. A budget-maximizing Forest Service will then require purchasers to build many roads, thus gaining access to otherwise inaccessible parts of the forest (figure 7.3).

A different strategy would be used in forests with low timber values. There, purchaser credits would limit the Forest Service's ability to obtain the maximum possible amount of revenues in K-V, BD, and TSS. The Forest Service would therefore minimize purchaser credits, either by building roads at a slower pace or by seeking appropriated road construction funds to substitute for purchaser credits.

Symptoms of Budget Maximization

The Brush Disposal Act, Knutson-Vandenberg Act, and other well-intentioned laws actually give the Forest Service an incentive to sell below-cost and negatively valued timber. A budget-maximizing Forest Service would retain as large a share of timber receipts as possible in the form of K-V deposits and deposits to other funds. The Forest Service can do this in several ways:

Figure 7.3 Alternative Road Construction Strategies

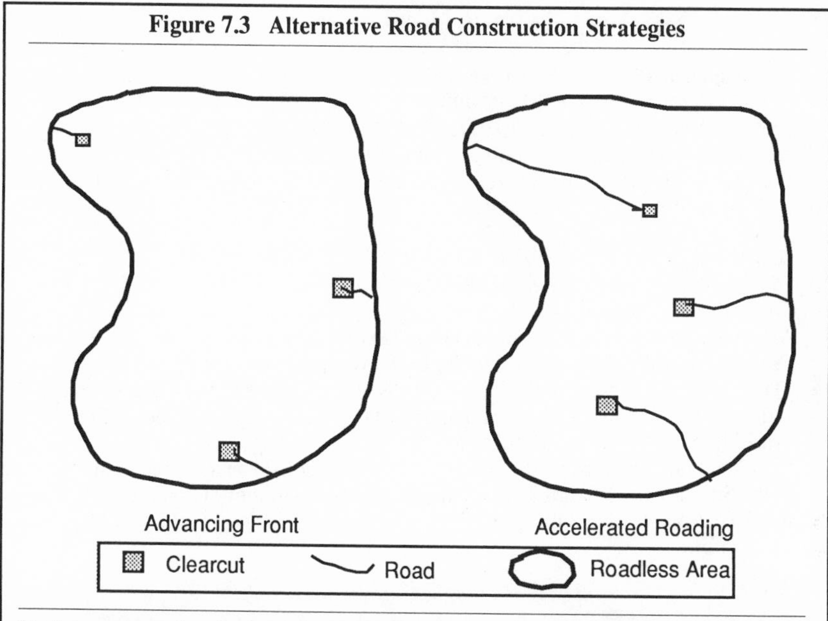

Under an advancing-front road strategy, roads are built a short distance into roadless areas to reach nearby stands of timber. Most of the roadless area remains intact. Under an accelerated road policy, roads are built deep into the heart of roadless areas so that later sales can salvage timber along the roadsides.

1. Cross-subsidize timber sales to increase marginal revenues and, in turn, the volume of timber it can sell and from which it can collect K-V, BD, and TSS funds.
2. Use timber management techniques that impose high reforestation costs and vary K-V collections according to the value of the timber sold.
3. In forests with high timber values, apply surplus income to purchaser credits to rapidly expand the road network.
4. In forests with low timber values, attempt to minimize purchaser road costs by building roads at a slow pace and by substituting appropriated road funds for purchaser credits.

A specific test to distinguish a budget maximizer from a timber primacist would be the presence of mechanisms to protect the agency's budget from the revenue effects of negatively valued timber. The existence of such mechanisms, which reduces the potential amount of timber that can be sold, would suggest that the agency is a budget maximizer, not a timber primacist. Chapter 8 will review recent Forest Service timber sales and budgets to see if these strategies are actually being used by forest managers.

Notes

1. Act of June 4, 1897 (provision repealed).
2. Act of August 11, 1916 (16 USC 490).
3. Knutson-Vandenberg Act of 1930 (16 USC 576-576b).
4. National Forest Roads and Trails Act of 1964, section 4 (16 USC 535).
5. National Forest Management Act (NFMA) of 1976, section 14(h) (16 USC 472a).
6. NFMA, section 14(a) (16 USC 472(a)).
7. NFMA, section 18 (16 USC 576b).
8. Senate Report No. 90-893, 94th Congress, second session, p. 20.
9. 36 *Code of Federal Regulations* 223.4(a).
10. *Forest Service Manual* 2421.3.
11. Alfred A. Weiner, *The Forest Service Timber Appraisal System: A Historical Perspective, 1891-1981* (Washington, DC: Forest Service, 1982), p. 12.
12. "Timber Sales Program of the U.S. Forest Service," unpublished paper on file at the office of the Surveys and Investigations Staff, House Appropriations Committee, Washington, DC, February, 1984, p. 46.
13. "Timber Sales Process of U.S. Forest Service," unpublished paper on file at the office of the Surveys and Investigations Staff, House Appropriations Committee, Washington, DC, April 1982, p. 40.
14. Ibid., p. iii.

Chapter Eight
Testing the Hypothesis

A Forest Service official told me one day that he had seventeen people working on his staff. "I sat down once and figured out that, if I had to, I could get along with only nine," he revealed. "Eight of them are just pushing paper around.

"But," he went on, "I realized that a government employee who has a staff of only nine doesn't get paid as much as one who has a staff of seventeen. So I decided not to do anything about it."

A close look at national forest timber sales shows that the Forest Service, like this employee, is sensitive to its budget. National forests often do retain all or most timber receipts in the form of purchaser credits or K-V deposits. To do so, they use all of the techniques described in chapter 7. In addition, the Forest Service clearly

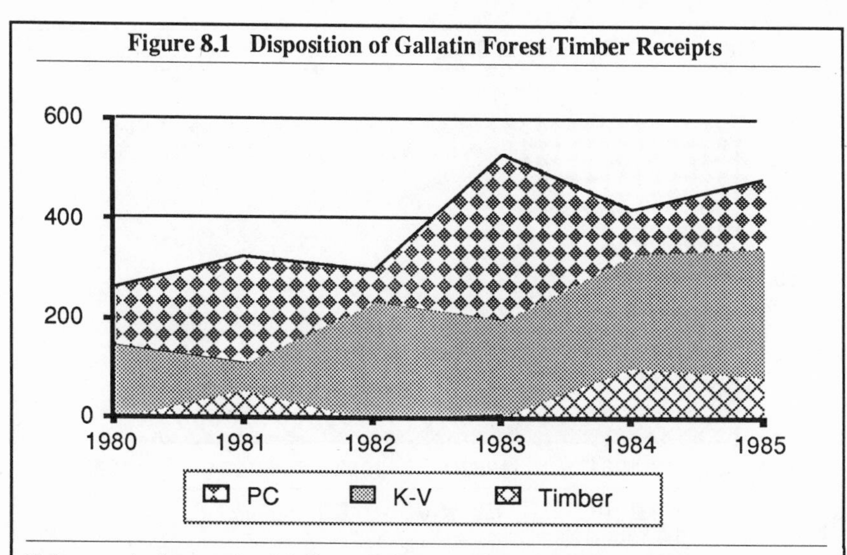

Figure 8.1 Disposition of Gallatin Forest Timber Receipts

Values are in thousands of dollars. Only a small portion of Gallatin Forest timber receipts have gone into the timber account in recent years. The remainder has been used to build purchaser-credit roads (PC) or retained for future reforestation activities (K-V).
Source: Forest Service Region 1 collections statements for years indicated.

protects its own budget from sales of negatively valued timber, a practice that firmly distinguishes it from a timber primacist.

Negative Net Receipts

Forest Service timber sale collections are recorded in three accounts: K-V, purchaser credits, and timber. Records are kept for these accounts for the benefit of counties which are guaranteed a 25 percent share of the total of the three accounts. These records show that many forests in the Rocky Mountains and at least a few national forests in each Forest Service region often retain almost all timber receipts as either K-V or purchaser credits.

For example, the Gallatin National Forest collected no receipts in its timber account in 1982, and less than 2 percent of its timber receipts went into the timber account in 1983 (figure 8.1).[1] All remaining receipts were either collected in the form of purchaser-constructed roads or deposited into the K-V trust fund.

When such a large percentage of receipts is retained by the Forest Service, the Treasury collects "negative receipts" net of county payments. Since the Treasury must pay 25 percent of the total of the three accounts to counties, when the timber account is less than 25 percent of the total, the Treasury must allocate other money to counties. The receipts, net of county payments, are negative. The Bridger-Teton's timber account, for example, has not exceeded 25 percent of its total timber collections in any of the past six years (figure 8.2).[2] Many other forests in the

Figure 8.2 Disposition of Bridger-Teton Forest Timber Receipts

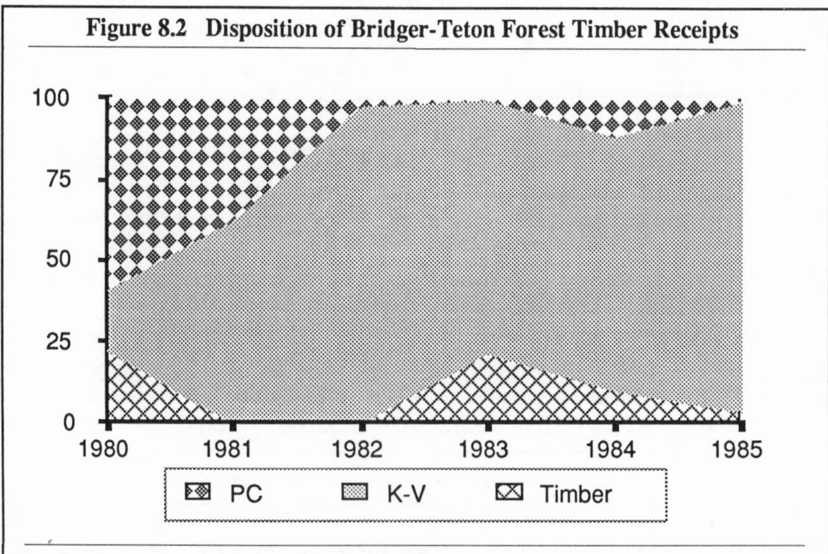

This figure shows returns on a percentage basis while figure 8.1 indicated actual dollar returns.
Source: Forest Service Region 4 collections statements for years indicated.

Figure 8.3 Disposition of 1985 Timber Receipts for Seven Forests

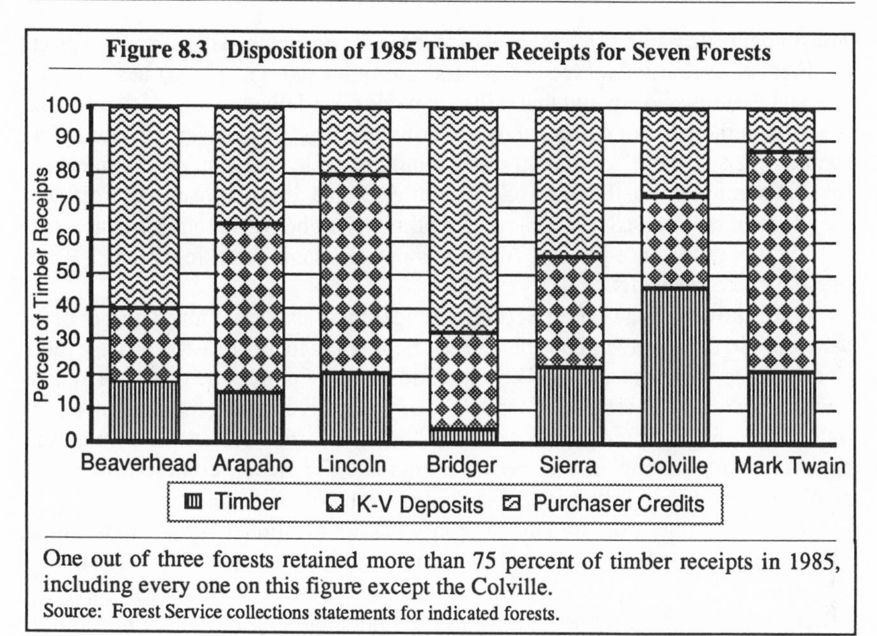

One out of three forests retained more than 75 percent of timber receipts in 1985, including every one on this figure except the Colville.
Source: Forest Service collections statements for indicated forests.

Rocky Mountains, where timber values are particularly low, have collected an average of less than 25 percent of their receipts in their timber accounts since 1980.

In 1985, nearly one out of every three forests, including at least one forest in every region, collected less than 25 percent of timber receipts in its timber account (figure 8.3). In 1980, when timber prices were higher, only about one out of seven forests did so, but there was still at least one such forest in every region. Out of the 140 different national forest units for which timber receipts are recorded, the timber accounts for 44 forests were smaller than 25 percent of total timber collections (which include K-V and purchaser credits) in 1985 — meaning that they collected negative net receipts. Nineteen did so in 1980, a considerably better year for the timber industry.

Most of the forests that return negative net receipts are in Regions 1 through 4, the Rocky Mountain and Intermountain regions. Forests in Oregon, Washington, northern California, and the deep South were least likely to return negative net receipts to the Treasury. This may be because the timber in these areas is so valuable that forest managers are unable to justify keeping most of the receipts for reforestation, brush disposal, and other activities.

Cross-Subsidization

Forest Service timber sale appraisals demonstrate that cross-subsidization is an extremely common practice in national forest timber sales. Most appraisals estimate the value of each species of timber in a sale to a purchaser of average efficiency.

The timber is then sold to the highest bidder but for no less than the appraised price.

Frequently, individual species in a sale are appraised at negative rates. In other words, the estimated amount that mills are paying for delivered logs of that species is less than the costs of cutting and delivering the logs. If the average value of all species in a sale were negative, the sale would be called a *deficit sale* — meaning that the purchaser, not the Forest Service, is expected to lose money. Deficit sales may occur if the Forest Service has underestimated timber values and purchasers are willing to pay the base price of fifty cents per thousand board feet plus K-V and brush disposal deposits.

Often, however, the Forest Service includes high- and low-valued timber in the same sale so that the average value of the sale is equal to, or close to, the base rates. For example, many sales in the West include stands of ponderosa pine — a highly-valued species — and stands of Douglas-fir, white fir, and lodgepole pine — low-valued species that often grow together in "mixed conifer" stands. To compensate purchasers for having to pay positive amounts for mixed conifer timber, the minimum bid price of the ponderosa is reduced below its appraised value.

The Dun timber sale, sold by Oregon's Malheur National Forest in December 1982, is a good example. Two million board feet of ponderosa pine in this sale was estimated to be worth about $53 per thousand board feet to a purchaser of average efficiency (table 8.1). Other species in the sale, including Douglas-fir, white fir, larch, and lodgepole pine, totaled 2.3 million board feet, but their value was estimated to be minus $30 to minus $63 per thousand board feet.[3]

Base rates for the species — meaning reforestation costs plus $0.50 per thousand board feet — ranged from $1 for lodgepole pine to $10 for ponderosa pine. Appraisers adjusted the lodgepole and other negatively valued species prices to base rates, and, to compensate purchasers, adjusted the ponderosa pine price down to $10.

Table 8.1 Dun Timber Sale Appraisal

	Ponderosa Pine	*Douglas-Fir*	*Western Larch*	*White Fir*	*Lodgepole Pine*
Volume (MBF)	1,960	1,050	820	370	30
Value to Mills	$415	$319	$325	$329	$206
Purchaser Costs	362	366	355	371	269
Indicated Advertised Rates	53	-47	-30	-42	-63
Base Rates	10	6	6	1	1
Adjustments to Base Rates	-43	53	36	43	64
Advertised Price	10	6	6	1	1
High Bid Price	$41	$6	$6	$1	$1

Ponderosa pine, which could have been sold for at least $53 per thousand board feet, was offered for only $10 per thousand. Although bidding raised the price of the ponderosa to $41 per thousand board feet, the sale returned almost $12,000 less than if the ponderosa had been sold alone.

Source: Forest Service timber sale report (form 2400-17) for Dun Timber Sale, Malheur National Forest, John Day, OR, December, 1982.

Table 8.2 Timber Appraised at Negative Rates in 1983

Region	*Total Volume (MMBF)*	*Volume Losing (MMBF)*	*Percent Volume Losing*
1	929	690	74
2	268	108	40
3	356	60	17
4	218	122	56
5	1,711	908	53
6	4,643	1,416	30
10	85	53	62
Total	8,210	3,357	41

Much of the timber appraised at negative rates was sold in cross-subsidized sales.
Source: Randal O'Toole, "Cross-Subsidies: The Hidden Subsidy," *Forest Planning* 5(2):15-17.

Three companies bid on the sale, sending the price of ponderosa pine to $41 per thousand board feet. The remaining species were sold for base rates, and the total sale price was $92,000. However, if the ponderosa pine alone had been sold for the "indicated advertised rate" of $53 per thousand board feet, the total sale price would have been nearly $104,000 — close to $12,000 more than the actual sale price.

The cross-subsidization in the Dun sale is not rare. Table 8.2 shows that 40 percent of the timber sold in fiscal year 1983, the year of the Dun sale, was cross-subsidized by other timber.[4] If negatively valued timber had not been included in sales, the Forest Service could have received bids of at least $40 million more than it did receive, and possibly as much as $146 million more (table 8.3).

Table 8.2 also indicates that the largest volumes of cross-subsidized timber were

Table 8.3 Revenue Losses Due to 1983 Cross-Subsidized Sales

Region	*Lost Receipts*	*Negative Appraisal*
1	$11,787	$54,569
2	32	1,734
3	64	2,562
4	719	7,635
5	9,898	35,431
6	17,210	40,055
10	2,622	4,304
Total	$42,332	$146,290

The "lost-receipts" method of calculating revenue losses apportions purchaser over-bids (the amount by which purchasers' bids exceeded the appraised value) to all species in the sales. The "negative-appraisal" method apportions overbids only to positively valued timber. The true loss is probably somewhere between these extremes. However, some methods of cross-subsidizing timber are not counted in this table. Dollars are shown in thousands.
Source: Randal O'Toole, "Cross-Subsidies: The Hidden Subsidy," *Forest Planning*, 5(2):15-17.

in Regions 1 (Montana and northern Idaho), 5 (California), and 6 (Oregon and Washington). Almost no timber was cross-subsidized in Regions 2 (Colorado and eastern Wyoming), 10 (Alaska), or other regions identified as having sales below cost — simply because there is no timber valuable enough in those states to compensate for negatively valued timber.

Cross-subsidies also take place between stands on gentle and steep slopes. Steep slopes require more costly logging techniques, so purchasers are reluctant to buy timber on steep slopes. In Arizona and New Mexico, the Forest Service is trying to overcome such reluctance by promising to include gently sloped timberland in every sale with steep slopes.

Thirteen such sales have recently taken place in New Mexico national forests. The steeply sloped portions of all but two of these sales lost money. Logging costs on the steep slopes averaged about $75 per thousand board feet, while on the gentle slopes those costs averaged about $30 per thousand board feet. Bid prices averaged only $10 per thousand board feet, which means that if the gentle slopes had been sold alone, they would have received bids of nearly $35 per thousand board feet. The money-losing, steeply sloped portions of these timber sales cost about $350,000 to arrange and administer. Because these units were included in the sales, purchasers bid an estimated $568,000 less for the sales.[5] The total cost to taxpayers, then, was well over $900,000, but the Forest Service gained almost $100,000 in K-V deposits.

Forest managers are often encouraged to cross-subsidize timber sales to "make sales attractive to industry." For example, in an August 1984 memo, the chief of the Forest Service encouraged regional foresters to "improve the economic viability of timber sales."[6] The Region 4 director of timber management responded by directing forests in Utah, southern Idaho, Nevada, and western Wyoming to "include high value species into sales with a significant volume of low value species."[7]

A more recent memo from Region 4 admits that Forest Service officials are aware of cross-subsidization. Negatively valued timber, the memo says, cannot be sold in isolation. "It can be sold, however," says the memo, "in combination with positively valued timber as long as the total value of the sale offering is positive. Repackaging of sales to produce more economically viable offerings is a common practice."[8] One of the authors of this memo, the Region 4 regional economist, notes elsewhere that the "timber industry often refers to [negatively valued units as] 'punishment units.' "[9]

Maximizing Reforestation Costs

A review of reports for sales in 1983 indicates that K-V deposits charged by individual forests are often correlated to the value of the species being sold. Species that are worth more on a given forest are charged higher deposits. The Dun sale illustrates this: Table 8.1 shows that the K-V deposit for ponderosa pine on this sale was $9.50 (base rates minus $0.50) per thousand board feet, while it was only $5.50 for Douglas-fir and larch, and only $0.50 for white fir and lodgepole. These are typical of base rates for ponderosa pine and mixed conifer forests, where ponderosa is the most valuable species by far and Douglas-fir and larch are considered to be

Figure 8.4 Disposition of Bridger-Teton Forests Receipts

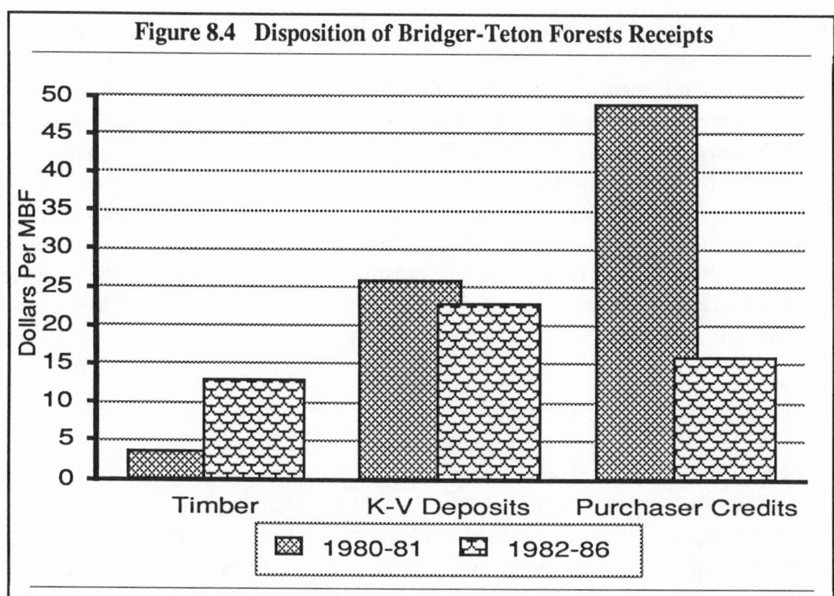

Portion of receipts alloted to timber, K-V deposits, and purchaser credits in two time periods in dollars per thousand board feet for the Bridger-Teton National Forests. When timber prices fell after 1981, Bridger-Teton managers apparently responded by reducing the amounts of K-V and purchaser credits purchasers would have to pay. By making the largest reductions in purchaser credits, managers were able to protect most, but not all, of their K-V income.

Source: CHEC, "Review of the Bridger-Teton Forests Plan" (Eugene, OR: CHEC, 1986), p. 4.

more valuable than white fir and lodgepole. (Douglas-fir has a lower value than white fir in the Dun sale, possibly due to smaller tree diameters.)

To some degree, this variation in K-V deposits makes economic sense. More money should be spent reforesting the most valuable species. But it also makes budgetary sense for the Forest Service, which can charge more for the most valuable species.

Forests that have highly valued timber tend to charge more than forests with marginal timber. K-V reforestation charges in Regions 1 and 6 range from $5 to $15 and occasionally as high as $30 per thousand board feet. This is significantly more than forests in Colorado (part of Region 2), where timber values are much lower and K-V deposits rarely exceed $2.50 per thousand board feet. Timber values are higher in South Dakota's Black Hills National Forest, which produces the most valuable timber in Region 2. There the K-V reforestation deposits are typically about $7.50 per thousand board feet.

A memo relating to the "Headache Heaven" sale in the Payette National Forest (Idaho) reveals managers' reasoning behind K-V charges. The regional base rate for the sale would have been $6 per thousand board feet. The memo notes that the forest used the "Forest standard rate," or $8 per thousand, instead. "The reasoning was that the amount was higher and we were trying to reclaim the most dollars," explains the memo.[10]

In marginal forests, timber values are so low that increases in K-V deposits or purchaser credits can make the difference between whether or not a sale occurs.

In such forests, changes in timber values over time are reflected by changes in K-V deposits. For example, when timber prices were fairly high in 1980 and 1981, the Bridger-Teton Forest retained an average of $25.37 per thousand board feet in K-V deposits. Between 1982 and 1986, after prices fell, the forest retained only $22.90 (figure 8.4).[11] Purchaser credits, meanwhile, fell from $49.17 to $16.09 per thousand board feet, suggesting that managers would rather lose purchaser credit income than K-V income.

Discretionary Expenditures

Once K-V, brush disposal, and timber salvage sale deposits are collected, they can be spent on their respective programs. Yet a large share of these funds is actually spent on other activities. Congressional appropriations include money for administrative overhead, but no overhead money is appropriated for the K-V, brush disposal, and timber salvage sale funds. Each level of the Forest Service — the Washington office, the regional offices, and the forest supervisors' offices — therefore retains a share of these funds for overhead and "indirect expenses."

The exact share varies from year to year and from forest to forest (table 8.4). Typically, for each dollar spent on reforestation or brush control by a district ranger, the Washington office might collect $0.04. Regional offices take a larger share, usually around $0.08 to $0.10, and they also take a share of timber salvage sale expenditures. Supervisors' offices usually retain $0.15 to $0.25. As a result, an average of one-fourth of the money that is supposedly spent on reforestation, brush disposal, and timber salvage sales is actually spent by higher levels of the bureaucracy. National forest expenditures from these funds exceeded $200 million in 1986, of which $50 million was spent on overhead.

The overhead money and indirect costs cannot be spent by the upper echelons until the districts actually spend money in the field. Because budgetary planning

Table 8.4 Portion of Timber Collections Spent on Overhead

Forest	*Knutson-Vandenberg*	*Brush Disposal*	*Salvage Sale Fund*
Gallatin (MT)	26	27	24
Klamath (CA)	25	39	36
George Washington (VA)	45	n.a.	23
Gila (NM)	35	35	17
Shoshone (WY)	24	24	23
Siskiyou (OR)	32	37	27

Numbers represent the percent of the total expenditure from each fund that is spent by the Forest Service on general administration and indirect expenses.
Source: *Forest Service Manual* 6522.57 for the indicated forests.

begins several years in advance of spending, this can sometimes create a financial squeeze. For example, budgetary planning for fiscal year 1982 began in 1979. The timber industry was doing well in 1979, and planners estimated that many acres would need reforestation and brush disposal treatments in 1982. The overhead and indirect costs that each office would spend in that year were based on these estimated reforestation and brush disposal objectives.

By 1982, however, the industry was in the midst of a severe depression. Little timber had been cut in 1981, so there was little need for reforestation and brush disposal in 1982. But the higher offices were still expecting to spend money budgeted for overhead and indirect costs, so they placed pressure on district officers to meet their reforestation and brush disposal "targets."

Unnecessary Work and Falsified Data

In 1985, a Forest Service employee in California was concerned about the effects of these targets on district activities. With the approval of the Forest Service, Cherry DuLaney, of the Tahoe National Forest, surveyed district silviculturists to ask how they responded to the targets between 1980 and 1984. Her results were startling.

Of silviculturists responding to the survey, 43 percent reported that some of the herbicide spraying they did in the past five years was not necessary or was done at the wrong time of year and would have no effect (table 8.5). They sprayed only to meet targets. DuLaney estimated that 15 percent of all herbicide spraying in those years was not needed. A similar percentage of silviculturists reported that some of the precommercial thinning they did was unnecessary. Twenty percent reported that some of the site preparation and reforestation was either unnecessary or ill-timed. In addition, some silviculturists admitted to falsifying reforestation and

Table 8.5 Strategies Used to Meet Silvicultural Targets

Operation	*Poor Work (1)*	*Falsified Data (2)*	*Cost (3)*
Site Preparation	20	23	$549
Planting	20	6	1089
Herbicide Release	43	4	1075
Precommercial Thinning	42	8	649
Total			$3,362

Region 5 (California) silviculturalists report that they either did work which was unnecessary, or, in their judgment, would be unsuccessful, or they falsified data to meet reforestation and related targets. The Forest Service employee who prepared the survey estimated that this work cost the agency more than $3 million during the five-year period covered by the survey.

1) Percentage reporting unneeded or ill-timed work to meet targets
2) Percentage reporting falsification of data to meet targets
3) Estimated Region-wide cost ($1,000s) of unneeded work

Source: "The Effects of Annual Targets and Budgets on the Quality of Silvicultural Practices in Region 5," unpublished memo on file in the Region 5 Forest Service office, San Francisco, February 1985, pp. 1-2.

other records to show that they met targets.

DuLaney estimated that unnecessary, ill-timed, or falsified work cost Region 5 forests nearly $3.4 million between 1980 and 1984. She concluded that targets should be replaced by some other measure of an employee's success.

Given the budgeting process described here, an even flow of accomplishments is needed to maintain a stable organization. If trees are not planted in California, a Forest Service official in Washington may not get a desired pay increase. If herbicides are not sprayed in the Klamath National Forest, an administrator in San Francisco may lose his or her job. Silviculturists are pressured to meet reforestation and similar targets to maintain the flow of cash to higher levels of the bureaucracy.

Accelerated Road Construction

In 1960, USDA economist John Fedkiw suggested that "advanced" or "accelerated" road construction would allow the Forest Service to salvage dead or dying timber before the timber decayed. The value of this salvage, he suggested, would pay for the costs of accelerated road construction.[12] Studies by Forest Service research economists, however, found that the costs of accelerated road construction far outweighed the benefits. In a 1970 analysis of the possibility of doubling road construction rates in the Bureau of Land Management lands near Tillamook, Oregon, Con Schallau concluded that the "accelerated program would earn a minus 1.25 percent rate of return. Although advance roading would increase timber production and timber revenues, the additional revenues would not be enough to compensate for higher maintenance, timber sale administration, and interest charges."[13] After a 1972 study comparing normal road construction rates with four accelerated rates on the North Umpqua drainage in west central Oregon, Brian Payne concluded that "no accelerated rate of roadbuilding was economically justified."[14]

Nevertheless, after the National Forest Roads and Trails Act was passed in 1964 authorizing the use of purchaser credits, national forest managers in Oregon, Washington, and California began practicing accelerated road building. In the Rocky Mountains, where timber values were much lower, purchaser credits could not pay for accelerated road building, so road construction continued at a moderate pace. By the mid-1970s, when road controversies culminated in the Forest Service's two public roadless area review and evaluations (RARE I and RARE II), national forests in Oregon, Washington, and California had been heavily roaded. Those in the Rockies remained relatively roadless.

The average size of national forest roadless areas in Oregon in 1978, for example, was 17,000 acres, while in Montana the average was 28,000 acres. Out of a total of 177 roadless areas in Oregon, only 4 barely topped 100,000 acres in size, and 7 more exceeded 50,000 acres. In contrast, 8 out of 190 roadless areas in Montana were over 100,000 acres in size — 2 of them well over 200,000 acres — and 9 more exceeded 50,000 acres.[15] Clearly, Oregon forest managers had been building roads deep into undeveloped areas on an accelerated basis.

Appropriated Road Construction

In 1985, appropriated road funds contributed to the bulk of timber-related road construction in Regions 2, 3, and 4 — Rocky Mountain or Intermountain regions that are commonly believed to sell most timber below cost. In comparison, figure 8.5 shows that over 90 percent of timber-related road costs in the Pacific Northwest Region — which has the most valuable timber of any Forest Service region — were paid for with purchaser credits. Clearly, purchaser credits are used only when timber is valuable enough to give managers a surplus above the K-V costs.

Protecting the Budget

Although the Forest Service sells billions of board feet of negatively valued timber each year, it protects its own budget from the revenue effects of such sales. Each of the three main types of revenues — brush disposal deposits, reforestation deposits, and timber salvage sale deposits — is protected in a different way.

Brush disposal deposits are charged to the purchaser over and above the high bid price. Purchaser road costs cannot be credited against this deposit. Since the brush disposal deposit represents a premium over the minimum bid price, the fund is absolutely protected against revenue losses from negatively valued timber.

The Knutson-Vandenberg reforestation deposit is included in the minimum bid

Figure 8.5 1985 Road Construction Funding for Selected Regions

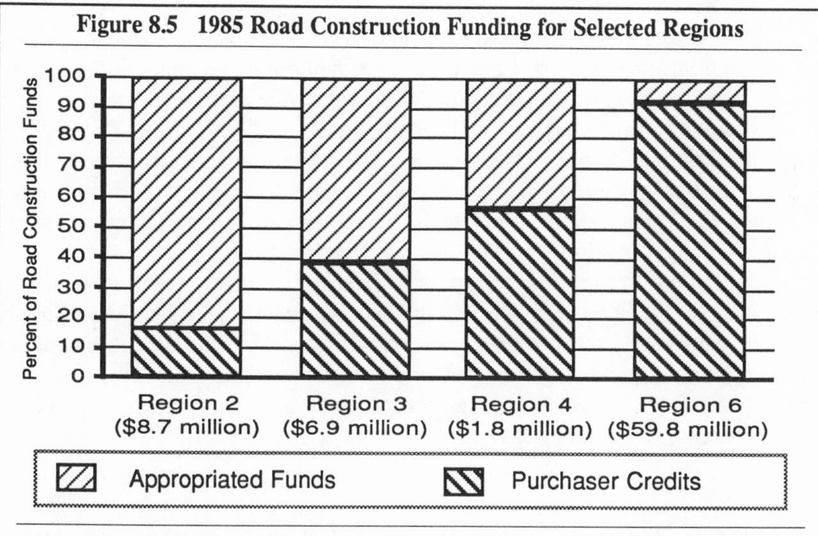

Appropriations paid for most road construction in Regions 2 and 3, where most timber is sold below cost. Purchaser credits paid for most road construction in Region 6, where most timber is sold above cost. Purchaser credits appear to be used only where timber is valuable enough to give managers a surplus over K-V costs.
Source: Forest Service 1985 budgets for the indicated regions.

price for the timber sale. The base rates, in fact, are equal to the the reforestation deposit plus $0.50 per thousand board feet. Purchaser road costs cannot be credited against this deposit. Thus, the reforestation deposit is also protected against revenue losses from sales of negatively valued timber as well as from deficit sales caused by high purchaser road costs.

Timber salvage sale deposits are not as clearly protected against low bid prices caused by negatively valued timber. Managers may plan to retain part of the revenues from a sale for the salvage sale fund, but this is not established as the minimum bid price. If the sale price is less than the amount that would have been retained, the fund will fall short. However, managers are then allowed to retain a premium on other salvage sales over what would ordinarily have been retained. The result is that individual salvage sales can be used to cross-subsidize one another, which may actually give managers added flexibility while not risking their budgets. In addition, when both K-V and timber salvage funds are retained from a sale, managers are allowed to keep all of the receipts — the Treasury is not even guaranteed the nominal $0.50 per thousand board feet which it collects from other sales.

The only part of the budget that is not protected is the portion of K-V funds that is used for wildlife habitat improvements and other nonreforestation activities. Although the K-V Act was passed in 1930, retention of funds for nonreforestation activities was not authorized until 1976. Since the timber division, which sets policies for K-V funds, does not retain funds for wildlife and recreation activities, it has little incentive to protect these funds. But other Forest Service divisions will probably urge that these funds receive the same protection as reforestation funds.

The Forest Service timber sale process provides perverse incentives to forest managers. Prices and costs fail to act as signals telling managers when the greatest net benefits are produced. Rather than equate marginal revenues with marginal costs, managers are encouraged to equate average revenues with the marginal costs of reforestation, brush disposal, and salvage sales (see figure 7.2). This practice leads to overproduction, below-cost sales, and sales of negatively valued timber. While negatively valued timber reduces revenues to counties and the federal Treasury, the Forest Service protects its own budget from such losses, indicating that it is a budget maximizer rather than a timber primacist.

This conclusion supports the hypothesis that budget maximization is a primary motivation for the Forest Service. A valid question is whether other Forest Service policies, particularly ones that seem economically inefficient, also fit the budget-maximization hypothesis. Chapters 9 and 10 will show that budget maximization can explain Forest Service policies regarding nondeclining flow, clearcutting, roadless areas, and other issues.

Notes

1. Gallatin National Forest collections statements for the years indicated, on file at Gallatin Forest supervisor's office, Bozeman, MT.
2. Bridger-Teton National Forests collections statements for the years indicated, on file at Bridger-Teton Forest supervisor's office, Jackson, WY.
3. Randal O'Toole, "Reviewing a Timber Sale Report," *Forest Planning* 4(6):8–10.
4. Randal O'Toole, "Cross-Subsidies: The Hidden Subsidy." *Forest Planning*

5(2):15–17.

5. CHEC, *Review of the Draft Carson Forest Plan* (Eugene, OR: CHEC, 1985), p. 11. CHEC, *Review of the Draft Cibola Forest Plan* (Eugene, OR: CHEC, 1985), p. 7; CHEC, *Review of the Draft Gila Forest Plan* (Eugene, OR: CHEC, 1985), p. 9; CHEC, *Review of the Draft Lincoln Forest Plan* (Eugene, OR: CHEC, 1985), p. 6.
6. "Need to Improve the Economic Viability of Timber Sales," memo from the chief to regional foresters, Washington, DC, 10 August 1984, 2 pp.
7. "Economic Viability of Timber Sales," memo from Region 4 to forest supervisors, Ogden, UT, 14 September 1983, 6 pp.
8. "Negative Stumpage Values," memo to the record by Region 1 and Region 4 economists, 10 June 1986, 4 pp.
9. David Iverson, "Sales Below Cost: Preliminary Notes Relating to Managerial Incentives in the Forest Service," 14 January 1985, memo on file in Forest Service office, Ogden, UT, p. 2.
10. Douglas Austin, "Headache Heaven Salvage Sale," memo from Council District Ranger to Payette Forest supervisor, 7 October 1982, Council, ID, 1 p.
11. Bridger-Teton National Forests collections statements for years indicated, on file at Bridger-Teton Forest supervisor's office, Jackson, WY.
12. John Fedkiw, "Advanced Roading for Increased Utilization in the Douglas-Fir Region," *Proceedings of the 51st Western Forestry Conference* (Portland, OR: Western Forestry and Conservation Association, 1960), pp. 64–68.
13. Con Schallau, *An Economic Analysis of Accelerated Road Construction on the Bureau of Land Management's Tillamook Resource Area* (Portland, OR: Forest Service, 1970), p. i.
14. Brian Payne, *Accelerated Roadbuilding on the North Umpqua — An Economic Analysis* (Portland, OR: Forest Service, 1972), p. iii.
15. USDA Forest Service, *Roadless Area Review and Evaluation Draft Environmental Statement* (Washington, DC: Forest Service, 1978), supplements for Oregon (Exhibit 2) and Montana (Appendix A).

Chapter Nine
Sustained Yield Maximizes the Budget

While reviewing the draft plan for one of the national forests near Yellowstone Park, I asked a planner about the rationale for the proposal. According to planning documents, far more people in the local communities depend for employment on forest recreation than on timber. Yet proposed timber sales in roadless areas would reduce the forest's recreation potential. The forest loses so much money on timber that each of the few timber-related jobs costs taxpayers tens of thousands of dollars per year. Why doesn't the Forest Service just stop selling timber?

"Organizational stability," was his answer. "The timber managers on this forest know that they would lose their jobs if this forest became a recreation forest. Although we might hire some more recreation managers, they would not be the same people who are now managing the timber. In addition, we probably need fewer recreation managers than the number of timber managers we now have."

Community stability has long been a proclaimed goal of the Forest Service. But, as this planner suggests, the Forest Service is more interested in its own stability than in that of the communities near its forests. This is well illustrated by the story of the Forest Service's policy of nondeclining flow.

Nondeclining flow — the policy of selling no more timber today than can ever be sold in the future — has been the subject of intense debate for several decades. The timber industry maintains that nondeclining flow wastes valuable resources.[1] Environmentalists insist that nondeclining flow protects community stability and other forest resources. For the most part, the Forest Service claimed to follow the policy long before it was first implemented under the name of "nondeclining even flow" in 1973.

But instead of considering nondeclining flow and community stability to be firm goals, the Forest Service appears to use these policies as tools to obtain larger budgets from Congress. Paradoxically, a policy of selling less than the maximum possible amount of timber can be used to increase the agency's budget. But once the goal of larger budgets is achieved, the tools are discarded: The nondeclining flow policy is so frequently and knowingly violated that the Forest Service's purpose in carrying it out must be something other than community stability.

Maximum Sustained Yield

Nondeclining flow goes hand in hand with a policy of not harvesting timber before it reaches its maximum average annual growth, or *culmination of mean annual increment* (CMAI). Together, nondeclining flow and CMAI are the two central components of the Forest Service's objective—incorporated into the Multiple-Use Sustained-Yield Act—of maximizing the sustained yield of timber harvests. These policies can be understood best by examining the way a stand of trees grows.

Timber yield tables show that stand volume does not increase in a straight line over time. In other words, an acre of 50-year-old trees would not be expected to exactly double its volume by the time it reaches 100. Instead, annual growth is fairly rapid at young ages and tapers off as the trees get older (see figure 9.1). The *mean annual increment* (MAI) of a stand is equal to the volume of the stand divided by its age. The MAI also reaches a peak and tapers off, but its peak takes place later than the growth curve. In calculus, the growth curve is the first derivative of the volume curve, while the mean annual increment curve is the second derivative.

Suppose that a forester has 100 acres to manage. If 1 acre is cut and reforested each year, then all the acres will have been cut after 100 years. After that time, the trees that will be cut each year will always be exactly 100 years old. This is called a 100-year *rotation*, and, when the trees being cut are always 100 years old, the 100-acre forest is called a *regulated forest*. Assuming that trees grow at the same rate on each of the acres, exactly the same volume will be cut each year by continuing

Figure 9.1 Volume and Growth of Medium Site Douglas-Fir

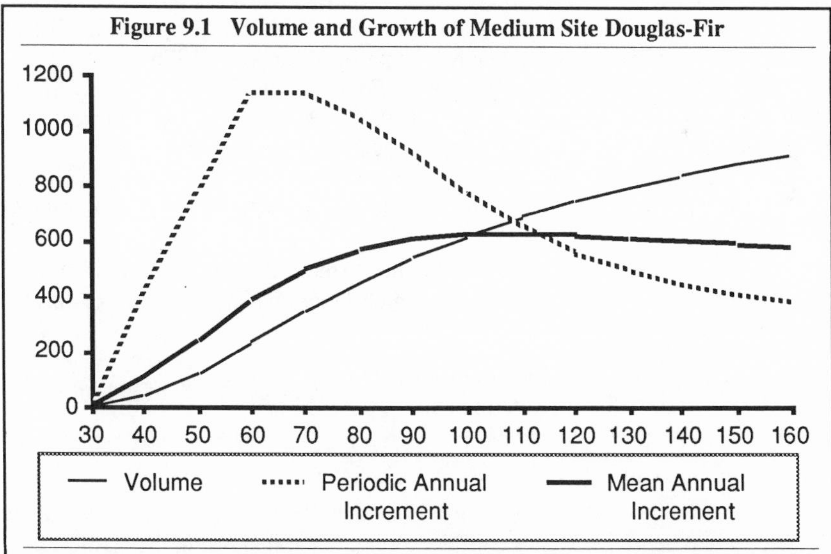

Volume is in hundreds of cubic feet per acre, while periodic and mean annual increment are in cubic feet per acre per year.

Source: Richard McArdle, *Yield of Douglas-Fir in the Pacific Northwest* (Washington, DC: USDA, 1930), p. 40.

to cut 1 acre per year. That volume will be equal to the average annual growth per acre times 100, the number of acres.

For the forest in figure 9.1, the mean annual increment at age 100 is 630 board feet per acre. This means that 63,000 board feet will be grown in 100 years. Thus, 63,000 board feet can be cut each year from the 100-acre regulated forest. What if 2 acres are cut each year? After the forest is regulated, the trees will all be age 50 when cut. The figure shows that the mean annual increment at age 50 is only 250 board feet per acre per year. This means that 12,500 board feet can be cut from each acre of 50-year-old timber or 25,000 board feet from 2 acres. In a regulated forest, then, cutting 2 acres each year can produce less volume than cutting 1 acre per year. Until the forest is regulated, of course, cutting 2 acres produces more volume than cutting 1 acre.

The mean annual increment in figure 9.1 reaches a peak of 635 board feet at age 110. If trees are cut at this age, then 69,850 board feet can be cut from each acre. Since about 0.909 acres would be cut each year, a total of 63,500 board feet would be cut per year.

Thus, the greatest possible amount of timber can be produced on a sustained basis if trees are harvested at the age where mean annual increment reaches a peak, or *culminates*. This is often called the *maximum sustained yield*. The CMAI for western forests ranges between 60 and 200 years, depending on forest type, land productivity, management intensity, and the minimum size of tree that is considered merchantable. Some highly productive southern forests reach CMAI sooner.

Many foresters have long argued for maximizing sustained yield — that is, harvesting stands of trees when and only when they reach culmination of mean annual increment. With the approval of the Forest Service, Congress established CMAI as the minimum harvest age for the national forests in 1976.

Having determined the rotation age, the Forest Service's goal is to convert unmanaged forests into regulated forests. If the forest is an old-growth forest, this means cutting down the existing timber. Early Forest Service officials believed that cut should be no more than growth. However, old-growth forests grow fairly slowly, and if cut were limited to growth, the forests might never achieve a regulated condition.

As an alternative, forest managers planned to harvest a volume equal to the old-growth volume divided by the rotation age plus the estimated growth rate of the old-growth:

$$\text{Annual harvest} = \frac{\text{Volume of old-growth}}{\text{Rotation age}} + \text{Annual growth of old-growth}$$

This is called *volume control*.

Some variation of volume control was used in Forest Service plans through the 1960s. Forest Service officials confidently predicted that, using these rules, they would always be able to cut as much timber in the future as they were cutting at that time. If this was their goal, however, their methods were seriously flawed.

Harvest Falldown Predicted

Most old-growth forests in the Pacific Northwest contained far more volume per acre than the second-growth stands were expected to grow by the age of CMAI. In forests where the rotation age was 100 years, for example, cutting 1 percent of the old-growth volume each year would produce far more wood than could be produced by second-growth forests. In 1969, the *Douglas-Fir Supply Study* found that harvest levels on most forests would "fall down" by as much as 30 percent when the old-growth was gone (figure 9.2).[2]

Although this seems easy enough to understand today, many people in the Forest Service were shocked when the *Douglas-Fir Supply Study* was published. The agency, which had been promising for so many years that it would never have to reduce harvest levels, suddenly found that it was running out of timber. If harvesting continued at current levels, timber harvests would decline in fifty to eighty years. To harvest only at sustained yield levels, however, meant dropping timber harvests immediately.

Forest Service scientists, however, thought they could find a way out of the dilemma. If second-growth timber could be grown faster, there would be less of a falldown in the future. Intensive management practices, such as thinnings, fertilization, and reforestation with genetically selected stock (sometimes called "supertrees"), could make up a large part of the difference between current and future yields. Reforestation of old brush fields would help make up the remainder.

Under a strict nondeclining flow policy, each acre of precommercial thinning

Figure 9.2 Harvest Levels on an Old-growth Surplus Forest

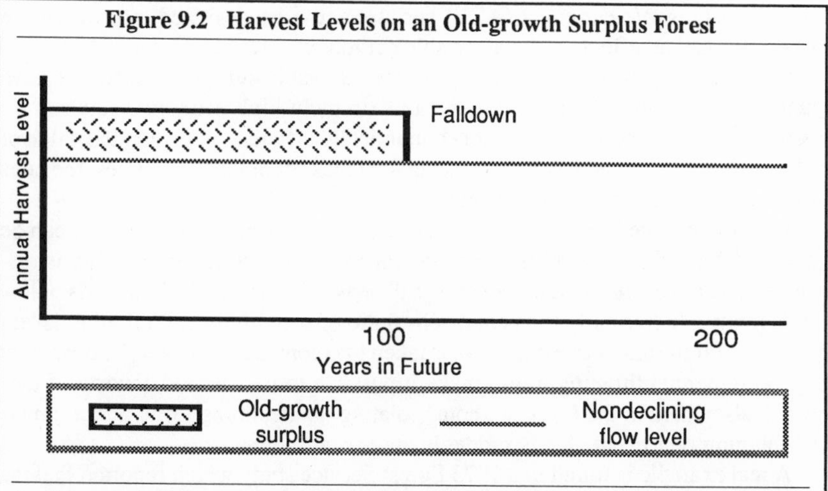

An old-growth surplus forest has more volume per acre in existing stands than managers expect to grow in second-growth stands, so harvests today can be larger than in the future. Under the nondeclining flow policy, however, harvests today cannot be larger than in the future. The timber industry calls the surplus that is lost under the nondeclining flow policy "even-flow waste."

Figure 9.3 Allowable Cut Effect on an Old-Growth Surplus Forest

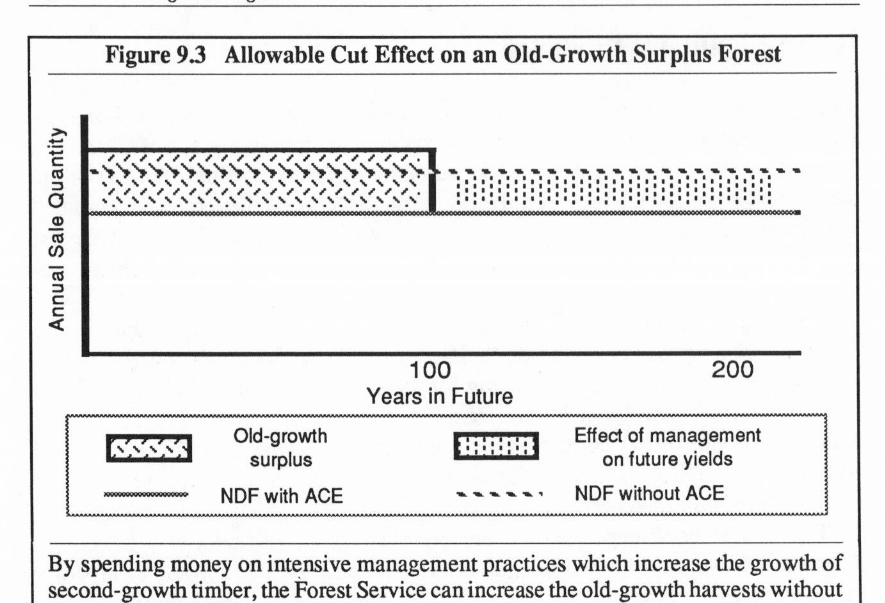

By spending money on intensive management practices which increase the growth of second-growth timber, the Forest Service can increase the old-growth harvests without exceeding nondeclining timber sale levels.

contributed to future timber growth and thus resulted in increased harvests of old-growth (see figure 9.3). Each brush field that was reforested meant more second-growth timber, allowing even more old-growth to be harvested today. This was called the *allowable cut effect* (ACE) because activities designed to grow future trees faster had an effect on the allowable cut today.

Because old-growth was so valuable, the allowable cut effect made intensive management practices appear to have an extremely high economic return. By themselves, these practices were not economically efficient, but when the value of this additional old-growth was counted against the cost of these practices, the rates of return were often over 50 percent.[3]

If $75 is invested in precommercial thinning on 1 acre, the timber that can be harvested from that acre might be more valuable. When the extra value is discounted back to the present, however, it will be worth less than $75 on lands of low productivity. In this case, the benefit-cost ratio of precommercial thinning is less than 1. But if the allowable cut effect is taken into consideration, a $75 investment in precommercial thinning might make it possible to harvest $150 worth of old-growth elsewhere in the forest without violating nondeclining flow. Precommercial thinning on low-site lands suddenly appears attractive.

A real example is found in a 1973 Forest Service study which reported that the present net value of thinning ponderosa pine stands in central Washington was negative using a 5 percent discount rate. However, when the allowable cut effect was considered, present net worth typically increased from negative $20 per acre to positive $60 per acre.[4] Such returns on investments in government activities were normally unknown. So, in 1973, the Forest Service decided to continue its sustained

Figure 9.4 Reducing Rotation Age on an Old-Growth Deficit Forest

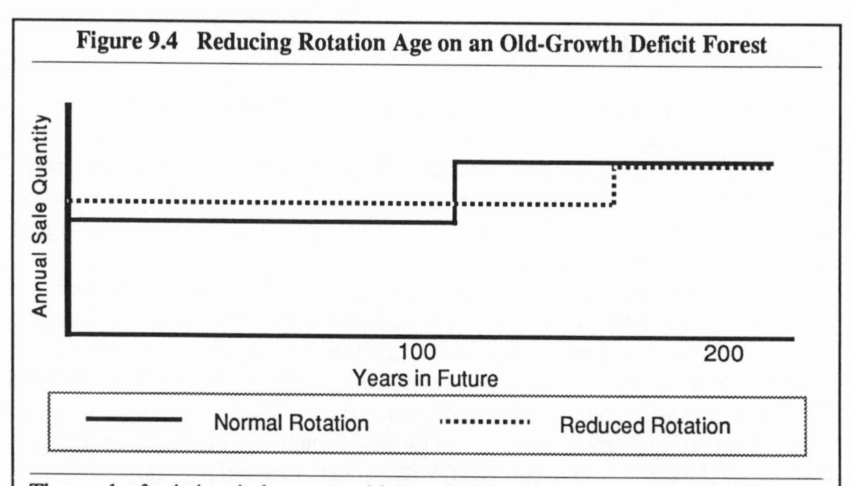

The stock of existing timber on an old-growth deficit forest is inadequate to maintain sales at the sustained yield capacity. Reducing the rotation age increases the volume of timber available for immediate harvest, allowing a higher nondeclining flow level. The trade-off is that a longer time is needed to reach the sustained yield capacity.

yield policy and required all future timber management plans to strictly conform to "nondeclining flow."

The timber industry considered this policy to be extremely wasteful. Since old-growth could be harvested rapidly today without ever falling below the sustained yield level, the difference in volume between high old-growth harvests and the nondeclining flow policy would be lost forever. The industry called this difference "even-flow waste."

Environmentalists supported the Forest Service's policy. At that time, old-growth was considered a "biological desert," but environmentalists believed some of it should be saved for posterity. If it were cut at a rapid rate, environmentalists would have less time to convince the Forest Service or Congress to set some aside. They were also concerned with the possibility that old-growth was a critical habitat for rare plants and wildlife. By cutting it slowly, there might be time to discover this fact before it was too late. Then land could be set aside from cutting to protect the habitat.

Nondeclining flow appears illogical from the viewpoint of timber primacy. It reduces the total volume of timber harvests over time, since the even-flow waste is never harvested. At the same time, it slows the conversion of old-growth to second-growth timber, whereas the mission of the timber primacist is often described as "converting the decadent old-growth to thrifty stands of rapidly growing trees." If the Forest Service were truly dominated by timber primacists, it is unlikely that it would have insisted on nondeclining flow.

Lower Harvests Equal More Money

At first glance, nondeclining flow also appears illogical from a budget-maximization view. Since harvests are reduced, the budget for sale preparation and administration is smaller. But using nondeclining flow and the allowable cut effect, the Forest Service could convince Congress to greatly increase funds for intensive management. In effect, the Forest Service could hold the allowable cut hostage to getting more money from Congress.

Dennis Teeguarden, a forest economist with the University of California at Berkeley, considered the high rates of return reported from the allowable cut effect to be suspicious. Teeguarden pointed out that the high returns were really due to harvests of existing timber — harvests that "could be realized even in the absence of" intensive management. Teeguarden termed the calculated rates of return "overestimated" and "misleading." "If new investments are evaluated on the basis of an ACE," he said, "poor resource allocation decisions will probably be made."[5] Yet forest managers were very willing to believe in the allowable cut effect because it confirmed their belief that timber management was worthwhile — and helped them convince Congress to fund timber management activities.

The framework under which Congress would fund intensive management was provided by the Forest and Rangeland Renewable Resources Planning Act—RPA. Passed the year after the Forest Service adopted the nondeclining flow policy, RPA directed the Forest Service to estimate its budgetary needs every five years, along with long-range estimates of the amount of timber and other resources that those budgets could produce. Congress would then fund national forest programs knowing what it would get in return.

From a budgetary viewpoint, this policy was extremely successful. Between 1970 and 1980, measured in constant dollars, the Region 6 budget for intensive management increased by 500 percent.[6] In Region 1, for example, appropriated funds for timber stand improvement — an important intensive management practice — increased by over 250 percent between 1970 and 1980.[7] Gains were not as spectacular in every region, but the total Forest Service budget for intensive management nearly doubled, in real terms, between 1970 and 1980.[8] These increases more than made up for any decline in sale preparation and administration dollars.

In addition, Congress authorized the Forest Service to use K-V funds for intensive management as well as reforestation in 1976. Partially as a result of this change, expenditures from the K-V fund expanded by over 50 percent between 1975 and 1980.[9] Most of this increase was in Regions 3, 5, and 6, where the most valuable western timber is found. Low timber prices in the other regions may have prevented forest managers from increasing K-V deposits.

Defending Nondeclining Flow

Today, forest planners use the term *old-growth surplus forest* for a forest that has so much old-growth that it can gain allowable cut effects from second-growth management. The term *old-growth* is used here in the traditional forester's sense of all trees over rotation age rather than in the biological sense, which requires that trees be much older.

Most forests in the West have old-growth, but not all are old-growth surplus forests. According to the Forest Service, many are *old-growth deficit forests*. In an old-growth deficit forest, the volume of an acre of second-growth timber at CMAI is more than the average volume per acre in the forest today. Thus, harvests can be expected to increase, not decrease, when the old-growth is gone and second-growth harvests begin. Eastern forests such as the Nantahala-Pisgah national forests (NC), which were mostly cut over in the past, are old-growth deficit forests. Other forests where natural growth has been slow due to overstocked stands may also be old-growth deficit forests. However, a forest may simply appear to be an old-growth deficit forest due to overly optimistic timber yield tables.

While intensive management leads to increases in today's allowable cut in an old-growth surplus forest, it does nothing for an old-growth deficit forest. Instead, the way to increase harvests in an old-growth deficit forest is to lower the rotation age. With a lower rotation age, more volume becomes available for immediate harvest, and more timber can be harvested (figure 9.4). The trade-off is that it will take longer for the forest to reach the sustained yield harvest level.

Most Bureau of Land Management (BLM) forests are old-growth deficit forests, and the BLM can increase harvests by reducing the minimum harvest age. The Forest Service cannot legally do this, however, due to the CMAI requirement in NFMA. At the time NFMA was passed, the term *old-growth deficit forests* did not exist, nor did the Forest Service suspect that most of its forests would fall into that category.

Despite assaults from a wide variety of sources, the Forest Service has staunchly defended the nondeclining flow policy since it was implemented in 1973. In 1979, the Carter administration directed the Forest Service to review the possibility of departing from nondeclining flow to reduce timber prices.[10] The Forest Service agreed to consider the possibility but, as Dennis Teeguarden notes, in reality they "completely ignored" Carter's directive.[11]

The Reagan administration made a stronger attempt to push the Forest Service into departures. Reagan appointed John Crowell, who was general counsel for Louisiana-Pacific, as assistant secretary of agriculture overseeing the Forest Service. Since Louisiana-Pacific buys more timber from the Forest Service than any other company, its executives have a particular interest in supporting departures from nondeclining flow.

Crowell quickly made it clear that he considered departures from nondeclining flow "not only desirable but absolutely necessary."[12] A secondary priority was to reduce minimum harvest ages below CMAI. Yet Crowell resigned in 1985 without a single forest departing from either policy, and many believed he left the job out of frustration over his inability to influence these and related issues.

Ignoring Nondeclining Flow

Despite this strong defense of nondeclining flow, the Forest Service has covertly ignored the policy. With little public disclosure, the agency has frequently set or proposed to set timber sale levels much higher than can be sustained in the future. It appears questionable whether the Forest Service truly believes in nondeclining flow as a means of protecting community stability — or anything else.

There are several ways of departing from nondeclining yield while appearing to follow the policy. One is to calculate timber sale levels using a different measure of wood than that used to sell the wood. A second is to combine two national forests, thereby gaining an increase in sales over the sum of the two separate forests. Sales can also be set at levels which are physically but not economically sustainable. Planners can make unrealistic assumptions about future timber yields or use unrealistically low CMAIs in setting rotation ages. Finally, a few forests have publicly proposed departures from nondeclining flow, yet ironically, these are not the forests that could most benefit from such departures. Each of these tactics will be discussed in the following sections.

Declining Sales in Board Feet

The amount of wood produced from a tree can be measured in several different ways. *Board foot* measures count only the amount of wood that is converted into lumber. *Cubic foot* measures count all the wood in a tree. A board foot is 1 inch by

Figure 9.5 Harvest Levels in Cubic and Board Feet

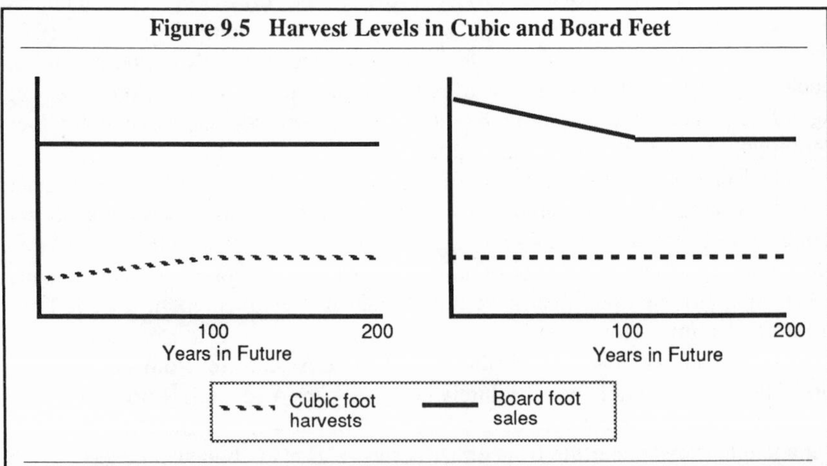

A second-growth forest generally will produce smaller diameter trees than an old-growth forest. As the old-growth is cut, the average diameter of trees will decline, and with it the number of board feet per cubic foot. If board foot harvests are even over time, cubic foot harvests will increase. If cubic foot harvests are even over time, board foot sales will decrease.

12 inches by 12 inches, while a cubic foot is 12 inches by 12 inches by 12 inches. But there are fewer than 12 board feet in a cubic foot: Because much of the wood in a tree cannot be converted to lumber, most forests have less than 6 board feet per cubic foot.

Moreover, the number of board feet per cubic foot is not the same for every tree. A greater proportion of a large tree than of a small tree can be made into lumber. An old-growth forest might produce 6 or 7 board feet for every cubic foot, while a second-growth forest may produce only 4 board feet per cubic foot. The part of the tree that is not converted to lumber may be used for paper, particle board, or other purposes, but these uses have a much lower value than lumber and plywood.

A forest that now consists primarily of large old-growth trees but that is being converted to a second-growth forest of smaller trees will experience a declining number of board feet for every cubic foot harvested. If the forest plans for non-declining yields in board feet, then the level of cubic foot harvests will increase over time (figure 9.5). If the forest plans for nondeclining yields in cubic feet, then the level of board foot harvests will decrease over time.[13]

Which measure should be used to calculate nondeclining yields? Lumber is far more valuable than the by-products that can be produced from the remainder of the tree. Nondeclining flows of board feet therefore more closely produce nondeclining flows of timber values. The timber industry has continually lobbied the Forest Service to sell most timber in board feet because timber purchasers mostly consist of lumber or plywood producers who cannot use the rest of the tree themselves.

Support for the notion that nondeclining yields should be calculated in board feet comes from the National Forest Management Act, which specifies that nonde-clining yield levels should be computed for "the *sale* of timber from each national forest" (emphasis added).[14] Since sales are in board feet, nondeclining yields should be calculated in board feet. Nevertheless, the Forest Service decided soon after the nondeclining yield policy was implemented to calculate timber sale levels in cubic feet rather than in board feet. In many forests, this decision means that board foot sales will decline by as much as 30 percent over time.

For example, the Stanislaus Forest projects an 11 percent increase in cubic foot outputs in the long term. However, the board foot-cubic foot ratio is likely to fall by close to 30 percent as old-growth harvests are replaced by harvests of smaller second-growth and yet smaller commercial thins. The result is that board foot sales will fall by 20 percent or more.[15] Yet the Stanislaus EIS fails to acknowledge this decline, instead claiming that board foot-cubic foot ratios will remain constant.[16]

Increased Harvests from Combining Forests

In 1947, the Forest Service made a unique arrangement with Simpson Timber Company on Washington's Olympic Peninsula. Most forestlands owned by Simpson had been completely cut over, and while the lands were mostly reforested, it would be many years before they would be merchantable again. The Olympic National Forest, meanwhile, had experienced very little timber cutting.

To protect the stability of the local communities of Shelton and McCleary,

Washington, the Forest Service agreed to combine an entire ranger district with Simpson Timber Company lands for the purpose of calculating the allowable cut. Simpson would have the exclusive right to buy timber from this ranger district. In exchange, Simpson would mill the timber it bought in Shelton and McCleary and give the Forest Service the authority to determine the rate at which Simpson could cut timber from Simpson's share of land in the arrangement.

When the allowable cut was calculated for the public and private land together, managers were surprised to find that it was 50 percent greater than the allowable cut for the two separately.[17] Today, the Forest Service land would be called an old-growth surplus forest, while Simpson land would be called an old-growth deficit forest. Combining the two resulted in increased cutting rates because the surplus forest could rely on the deficit forest to provide additional sustained yield capacity after the old-growth was gone.

Until 1960, the Forest Service frequently calculated the allowable cut for individual ranger districts. But, learning from the Shelton example, the Forest Service began combining ranger districts to increase rates of cutting. During the 1970s, proposals were made to combine several national forests, such as the Wallowa, Whitman, Malheur, and Umatilla, for the purpose of calculating the allowable cut.

Alarmed by the effects of increased timber cutting, environmentalists proposed that the National Forest Management Act require that allowable cuts be calculated for units of land no larger than ranger districts. Congress demurred from forcing the Forest Service to reverse its policies but agreed to limit the size of an allowable cut unit to individual national forests. As finally written, the law allows the combination of two national forests only if one has less than 200,000 acres of commercial forest land.[18]

To Congress, a national forest is an area of land that was "proclaimed" a forest by the president or an act of Congress. Although they are administered together, the Shasta and Trinity are really two separately proclaimed forests, as are the Wallowa and Whitman; the Kaniksu, Coeur d'Alene, and St. Joe; the Apache and Sitgreaves; the Grand Mesa, Uncompahgre, and Gunnison, and many other combinations. In all the cases just named, each proclaimed forest includes well over 200,000 acres of commercial forest land.

The Forest Service treats the forests as individually proclaimed forests when it calculates the returns to counties. Under this revenue-sharing formula, each county gets a share of forest revenues according to the number of acres of national forestland in the county.[19] The Whitman National Forest produces greater revenues than the Wallowa, and a county that includes acres from the Whitman would be opposed to sharing revenues with a county that includes Wallowa Forest acres.

The Forest Service is not as careful about making separate allowable cut calculations. In almost every case where two or more forests are administered together, forest planners have calculated nondeclining yields for all forests together. The environmental impact statement for the Idaho Panhandle Forests (which include most of the Kaniksu, Coeur d'Alene, and St. Joe forests) admitted that this would result in declining flows on all three forests at various times in the future.[20] Yet the Forest Service did not seem to regard this as a departure.

Figure 9.6 Harvest Levels in the Shasta and Trinity National Forests

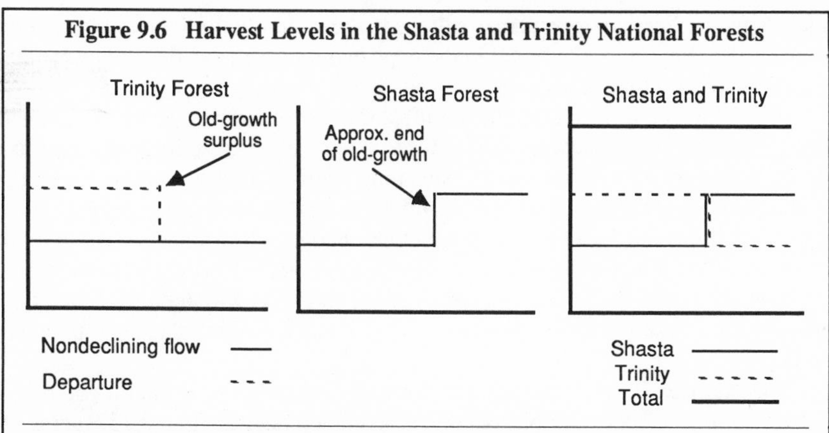

The Trinity is an old-growth surplus forest, so nondeclining sale levels will leave the old-growth surplus unharvested. The Shasta is an old-growth deficit forest, so harvests will increase after the old-growth has been replaced by second-growth. By combining the two forests, the total harvest level can be greater than the nondeclining harvest level of the two separately. Effectively, however, this represents a departure in the Trinity Forest, since harvests from that forest will decline over time.
Source: "Notes on the Shasta-Trinity 'Two Forest' Model," unpublished memo on file at the Shasta-Trinity Forests supervisor's office, Redding, CA, 22 September 1982.

The Shasta-Trinity Forest was an exception: Planners decided that the law required them to calculate sale levels for the two forests separately. In 1982, a draft forest plan based on these calculations was sent to the Washington office for review. Officials realized that the Shasta Forest was an old-growth deficit forest, while the Trinity was an old-growth surplus forest (figure 9.6). Planners were directed to combine the two forests and rewrite the plan. Ironically, the Washington Office appeared to be directing planners to violate the law to achieve the purpose — increased sale levels — the law was written to prevent.[21]

The revised Shasta-Trinity Plan and plans for other forests that consist of two or more proclaimed forests may produce nondeclining yields for the forests taken together. But nondeclining yields could just as easily be calculated for all the national forests in a region or in the entire nation. This would not ensure the protection of community stability. In most cases, individual mills tend to purchase timber mainly from one national forest. Although nondeclining flows are ensured on the combined forests, harvest levels on each individual forest will fluctuate, placing a potential hardship on the mills.

The Declining Even-Flow Effect

Timber sale levels that are physically sustainable may not be economically sustainable in many national forests. This can be true for either or both of two reasons. First, second-growth forest management requires a larger budget than old-growth management. Second, forests may sell the highest-quality timber today and defer poor-quality timber for the future in the hope that markets have developed for that timber — a hope that may not be realized.

Many forest plans will require significant increases in budgets as old-growth forests are converted to second-growth forests: Where old-growth management can be limited to road construction and sale preparation, second-growth management may involve expensive reforestation, thinning, and other management actions. Timber receipts may not keep pace with these budget needs since second-growth timber commands considerably lower prices than old-growth. In some cases, budgets must double in a few decades or timber harvests will decline. Since such budgets are unlikely if the timber does not pay its way, timber sale levels are likely to fall.

For example, one Oregon national forest calculated in 1977 that it could significantly boost its allowable cut by precommercial thinning 2,137 acres of forestland per year for ten years. To maintain the cut at the boosted level, the number of acres to be thinned would have to increase to 3,000 per year during the following ten years. By the third ten-year period, 12,109 acres would have to be thinned per year. Funds to maintain this level of management would have to increase by nearly six times in just twenty years. As a result, said Enoch Bell, a Forest Service economist, the "Forest Service policy of nondeclining flow could well become one of declining flow." Given a constant budget, the Forest Service calculated that harvests after twenty years would fall by 22 percent and would never return to the high level of the first ten years.[22]

The reason for this necessary increase in budget is simple. As Bell pointed out, "This national forest, like many others in the West, primarily contains trees beyond rotation age. This situation provides little opportunity for intensive management, since most silvicultural practices are applied to young stands. Acres harvested in the first decade, though, will be ready for stocking control treatments by the third decade. This time lag is what causes the large increase in treated acres required to maintain the nondeclining harvest level."[23]

Regardless of budget levels, Alan McQuillan has pointed out that forests that establish timber sale levels by asking FORPLAN to maximize present net value will often, in subsequent plans, produce declining levels of timber sales. This is because FORPLAN tends to include negatively valued timber in the timber base despite the maximum present net value objective. By cutting low-valued timber in the distant future, FORPLAN can cut more high-valued timber today.

The addition to present net value from cutting more high-value timber today easily outweighs the losses from selling money-losing timber in the future since those losses are heavily reduced by the discount rate. FORPLAN runs after the valuable timber has been cut will be less willing to include the money-losing land in the timber base, resulting in what McQuillan calls the "declining even-flow effect."[24] Forest planners in California have confirmed this effect by making consecutive FORPLAN runs, each one starting with the inventory of the previous run after ten years. Although each run projects nondeclining flows, subsequent decade calculations project nondeclining flows at lower levels than the previous decade.[25]

One of the more ludicrous examples of this lack of logic is found in the White River National Forest (CO). A significant share of the timber on this forest had been killed by insect epidemics. Planners included this volume in the FORPLAN model but noted that its value was very low. In order to meet minimum timber sale levels imposed by planners, FORPLAN included this timber in the suitable timber base — but because of its low value, FORPLAN postponed timber sales for 200 years. In 200 years, of course, dead trees will be little more than sawdust.[26]

Additional examples are found in the Cibola Forest Plan, which proposes to include timber on steep slopes in the suitable timber base. Local timber companies have resisted sales of such timber because it is costly to harvest. So, although the Cibola Forest includes over 52,000 acres of steep slopes in its timber base, it has no plans to sell any timber from steeply sloped lands for at least ten years. In the meantime, more timber can be cut on gentle slopes because the suitable timber base is larger. Sale levels may have to decline in the future if timber values never become high enough to justify selling timber from steep slopes.[27] In addition, 99 percent of the volume the Cibola Forest plans to sell in the next ten years is ponderosa pine, the most valuable species in the forest. To maintain timber sales, nearly 30 percent of the volume that will be sold in the future will consist of less valuable species, such as white fir.[28]

Planners in Region 4 have incorporated a "program viability constraint" into their FORPLAN models, which seeks to ensure that net timber receipts will not decline over time. Elsewhere, the Forest Service has done little to address the question of economic sustainability. Program viability constraints are not used anywhere else in the country, and as a result, many forest plans are proposing to defer harvests on low-value timberland until well in the future.

Questionable Yield Tables

To increase sale levels in spite of the nondeclining flow requirement, many forest plans propose practices that are expected to hasten the growth of second-growth stands. Since in an old-growth surplus forest nondeclining flows are limited by the growth rate of second-growth, any increase in that rate results in greater old-growth harvests. As noted earlier, this is called the allowable cut effect.

Projected increases in growth rates are sometimes valid, but often they are speculative or exaggerated. For example, most research indicates that precommercial thinning will increase the volume of merchantable timber at harvest age by increasing the diameter of the average tree. With larger diameters, the number of board feet per cubic foot is greater. However, research indicates that such increases are limited to about 20 or, at most, 30 percent of volume.[29]

Yet many forests use yield tables that project 50 percent or greater increases in volume due to precommercial thinning. There is no support for these claims. For example, to support yield tables that said its mixed conifer forests — forests consisting of Douglas-fir, true fir, and several pines — would grow 50 percent faster after precommercial thinning, the Flathead National Forest cited a study of lodgepole pine forests in northern Alberta.[30] Aside from the huge difference in climate between northern Alberta and northern Montana, lodgepole pine research is inapplicable to mixed conifer forests because lodgepole frequently stagnates if not thinned, whereas mixed conifer rarely does.

In the Pacific Northwest, many forest plans assume that the use of "genetically improved" seed for Douglas-fir plantations will increase volume by 10 to 15 percent.[31] The research basis for this is negligible. Few studies of Douglas-fir genetics have been completed, and planners arrived at the percentages by extrapolating from research on southern pines.

Roy Silen, the Forest Service's leading specialist on Douglas-fir genetics, is extremely cautious about projecting growth increases. He notes, "In our studies, progeny from the best parents in most stands produce about 20 percent more stem volumes than average." However, that volume is on a per tree basis rather than a per acre basis. "Virtually all instances in Douglas-fir come from plots of trees with room to grow," Silen continues. "Few experiments have yet reached the age or stage to express differences in family volume per unit area." However, he cites two studies that found that differences between "improved" stock and natural stock "dropped below statistical significance" when stands became crowded. Silen concludes that "the final word is still to come on whether genetic gains will be substantial without site enhancement."[32]

Forest plans are also beginning to project increases in growth from fertilization. Yet Silen also notes that "about a quarter" of the experiments with fertilization in the Northwest "have resulted in neutral or even negative response."[33] Projected growth gains from fertilization and genetics and large gains from thinning appear to be entirely speculative and undocumented by research.

Reducing Minimum Harvest Ages

As previously noted, nondeclining harvests from old-growth deficit forests can be increased by reducing minimum harvest ages. This increases the volume of "old-growth" (stands over rotation age) that is available for immediate sale. Since the volume of old-growth limits the nondeclining yields from an old-growth deficit forest, this in turn increases the nondeclining sale level.

Although the NFMA clearly states that stands of timber shall not be cut before they reach CMAI, many forest plans claim that CMAI will be reached at unrealistically young ages. Timber yield tables for California forest plans, which have never been reviewed by outside experts, predict that CMAI will take place at much younger ages than published yield tables that have been reviewed by scientific peers.[34]

The standard yield tables for Douglas-fir were prepared by Richard McArdle in 1931. McArdle, who later became chief of the Forest Service, measured thousands of trees and concluded that Douglas-fir stands would not reach CMAI before the age of 60. Moreover, if only trees above a certain diameter were counted — say, 7 or 12 inches — stands of average productivity would take 80 to 100 years to reach CMAI, and stands of lower productivity would take even longer.[35]

California national forest planners assume that the timber industry uses only trees over 11 inches in diameter for lumber. Counting only trees larger than this size, their yield tables predict that CMAI will be reached at the age of 50 — even in the lower site stands which McArdle predicted would not reach CMAI for 160 years. These unrealistically low CMAIs have the effect of increasing the volume of timber available for immediate harvest. Since Region 5 yield tables may also predict unrealistically high growth for second-growth stands, most California forests are considered old-growth deficit forests. Use of low CMAIs thereby increases the level of timber sales.

Departing from Nondeclining Yield in Old-Growth Deficit Forests

Timber industry representatives argue that departures can be made from nondeclining yield in an old-growth surplus forest with minimum disruption to the local economy (figure 9.2). The old-growth surplus will be harvested over a period of time, after which sales will fall to — but not below — the nondeclining sale level (also called the *sustained yield capacity*). Without the departure, the surplus will never be harvested. In the industry's view this is a waste — since all old-growth in the timber base is scheduled for eventual harvest, it will ultimately provide no environmental values. But if it is not harvested, it will also provide no timber values.

The industry is less supportive of departures from nondeclining flow in old-growth deficit forests. Departures in such forests require that future sale levels fall temporarily below the sustained yield capacity (figure 9.7). This would be far more disruptive to local communities.

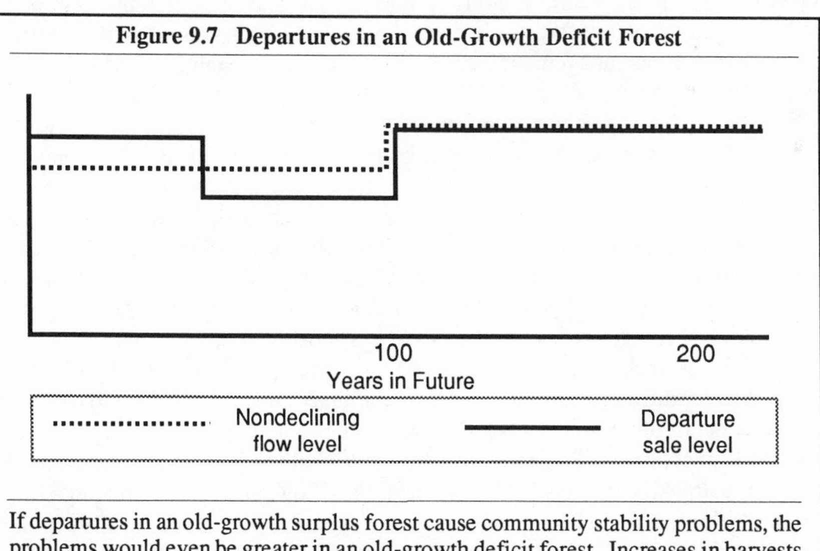

Figure 9.7 Departures in an Old-Growth Deficit Forest

If departures in an old-growth surplus forest cause community stability problems, the problems would even be greater in an old-growth deficit forest. Increases in harvests today require a trade-off of an equivalent decrease in harvests at some point in the future.

Of the more than 100 draft and final forest plans that have been published to date, only one has proposed a departure in an old-growth surplus forest. That plan, for the Klamath Forest (CA), was withdrawn, and its revised form is not expected to propose a departure. Other old-growth surplus forests, including the Wenatchee (WA), Siuslaw (OR), and Olympic (WA), have rejected departure alternatives.

Yet departures have been proposed by three old-growth deficit forests. Those plans, for the Deschutes (OR), Ochoco (OR), and Okanogan (WA) forests, admit that departures will require future sale levels to drop below the sustained yield capacity.[36] Local industry representatives have tended to oppose the departures.

The Forest Service appears willing to publicly support departures in old-growth deficit forests, where they will be most disruptive to local communities, but not in old-growth surplus forests, where they make the most economic sense. At the same time, the agency is covertly planning departures in both types of forests by using cubic foot rather than board foot measures, proposing economically unsustainable levels of sales, illegally combining forests for allowable cut calculations, and unrealistically speculating on future yields of timber — all of which indicate that nondeclining flow is not really a priority for the agency.

Score: Budget 1, Community Stability 0

From the viewpoint of timber sales, it appears that the Forest Service sacrificed little by adopting the nondeclining flow policy. Most forests in the West were thought to be old-growth surplus forests in 1970, but today only a few forests in northwest Oregon and western Washington have a surplus of old-growth. Forests in southwest Oregon, California, eastern Oregon and Washington, and elsewhere are now known to be old-growth deficit forests. In many cases, these forests have lost their surplus because intensive management increased the projected sustained yield capacity of the forests.

Whether these projections are realistic — and many analysts suspect they are not[37] — is irrelevant to a budget maximizing agency. The Forest Service now has the best of both worlds: large budgets for intensive management and high timber sale levels. Without continued support for intensive management from Congress, old-growth deficit forests will become old-growth surplus forests, and harvest levels will decline under the nondeclining flow policy. As long as the forests are old-growth deficit forests, however, there is no longer a need to use nondeclining flow. Thus, several old-growth deficit forests are proposing to depart from this policy.

If the Forest Service were controlled by timber primacists, it might have responded very differently to the *Douglas-Fir Supply Study*. According to Clary, the major concern of a primacist is the production of a maximum amount of wood to avoid a timber famine. Someone with this view would be horrified by the even-flow waste illustrated in figure 9.2. The nondeclining flow policy shows that maximum timber volume has not been the goal of the Forest Service. Because of the policy, harvests fell in the Gifford Pinchot National Forest and other forests, but their budgets increased.[38] Instead of maximizing timber production, the Forest Service took the logical steps to maximize its budget.

Notes

1. Wesley Rickard, *The Action Forest* (Salem, OR: State of Oregon, 1968), p. 2.
2. USDA Forest Service, *Douglas-Fir Supply Study* (Portland, OR: Forest Service, 1969), p. 14.
3. Dennis L. Schweitzer, Robert W. Sassaman, and Con H. Schallau, "Allowable Cut Effect: Some Physical and Economic Implications," *Journal of Forestry* 70(7):415–418.
4. Robert W. Sassaman, James W. Barrett, and Justin G. Smith, *Economics of Thinning Stagnated Ponderosa Pine Sapling Stands in the Pine-Grass Areas of Central Washington* (Portland, OR: USDA Forest Service, 1973), pp. 11–13.
5. Dennis E. Teeguarden, "The Allowable Cut Effect: A Comment," *Journal of Forestry* 77(4):224–226.
6. Region 6 budget statements for the years indicated.
7. Region 1 budget statements for the years indicated.
8. "OMB Historical Database," working paper on file at the Forest Service office, Washington, DC, 6 March 1986, p. 16.
9. Ibid., p. 32.
10. Alfred E. Kahn, "Statement on the President's Anti-Inflation Decision to Depart from Even-Flow Timber Harvest on Federal Lands," 11 June 1979, unpublished document on file at the Forest Service office, Washington, DC, p. 1.
11. Dennis Teeguarden, "A Public Corporation Model for Public Land Management," paper presented at the Conference on Politics vs. Policy: The Public Lands Dilemma, Utah State Unversity, Logan, UT, 23 April 1982, p. 12.
12. Joe Cone, "Interview with John Crowell," *Forest Planning* 2(1):10–13.
13. Roger Fight and Dennis Schweitzer, "What If We Calculate the Allowable Cut in Cubic Feet?" *Journal of Forestry* 77(11):701–702.
14. National Forest Management Act (NFMA) of 1976, section 11 (16 USC 1611).
15. CHEC, "Review of the Stanislaus National Forest Plan" (Eugene, OR: CHEC, 1985), p. 6.
16. USDA Forest Service, *Stanislaus Forest Plan Draft Environmental Impact Statement* (Sonora, CA: Forest Service, 1985), p. II-162.
17. Dennis L. Schweitzer et al., "Allowable Cut Effect."
18. NFMA, section 11 (16 USC 1611).
19. Act of May 23, 1908 (16 USC 500).
20. USDA Forest Service, *Idaho Panhandle National Forests Plan Draft Environmental Impact Statement* (Coeur d'Alene, ID: Forest Service, 1985), p. II-84–85.
21. "Notes on the Shasta-Trinity 'Two Forest' Model," memo on file at the Shasta-Trinity Forests supervisor's office, Redding, CA, 22 September 1982.
22. Enoch F. Bell, "Declining Harvest from Linking the Allowable Cut to the Budget on National Forests," *Journal of Forestry* 75(11):701–702.
23. Ibid. pp. 701–702.
24. Alan McQuillan, "The Declining Even Flow Effect — Non Sequitur of National Forest Planning," *Forest Science* 32(4): 960–972.
25. Interview with Klaus Barber, Region 5 systems analyst, 6 May 1986.
26. CHEC, "Review of the White River Draft Forest Plan" (Eugene, OR: CHEC, 1986), p. 15.
27. USDA Forest Service, *Cibola Forest Plan Proposed Amendment* (Albuquerque, NM: Forest Service, 1986), p. 17.
28. Ibid., pp. 131–33.
29. Donald L. Reukema and David Bruce. *Effects of Thinning on Yield of Douglas-Fir* (Portland, OR: Forest Service, 1977), p. 23.
30. Letter from Fred Hodgeboom, planning forester, Flathead National Forest, to Randal

O'Toole, 5 March 1985, pp. 3–4.

31. USDA Forest Service, *Siuslaw Forest Plan Draft Environmental Impact Statement* (Corvallis, OR: Forest Service, 1986), p. B-30.
32. Roy R. Silen, *Nitrogen, Corn, and Forest Genetics* (Portland, OR: Forest Service, 1982), p. 11.
33. Ibid., p. 13.
34. Randal O'Toole, *Analysis of Region 5 Timber Yield Tables* (Eugene, OR: CHEC, 1986), p. 13.
35. Richard McArdle, *The Yield of Douglas-Fir in the Pacific Northwest* (Washington, DC: Forest Service, 1930), p. 43.
36. USDA Forest Service, *Deschutes Forest Plan Draft Environmental Impact Statement* (Bend, OR: Forest Service, 1986), p. 45; USDA Forest Service, *Ochoco Forest Plan Draft Environmental Impact Statement* (Prineville, OR: Forest Service, 1986), p. 32; USDA Forest Service, *Okanogan Forest Plan Draft Environmental Impact Statement* (Okanogan, WA: Forest Service, 1986), p. II-9.
37. O'Toole, *Region 5 Timber Yield Tables*; and Randal O'Toole, *Timber Productivity of the Wallowa and Whitman National Forests* (Eugene, OR: CHEC, 1986).
38. USDA Forest Service, *Gifford Pinchot Timber Management Plan Final Environmental Statement* (Vancouver, WA: Forest Service, 1975), p. i.

Chapter Ten
More Budget-Maximizing Policies

Below-cost timber sales and nondeclining flow are not the only policies which are consistent with the budget-maximization hypothesis. Forest Service policies on many other controversial issues, such as clearcutting, wilderness, grazing, and recreation, can be explained if the agency is a budget maximizer. This chapter will present detailed discussions of clearcutting and wilderness and briefer discussions of grazing and recreation issues.

Clearcutting

The practice of removing all trees in a stand during a timber harvest has been controversial wherever it has been used. Clearcutting by private landowners was one of the main justifications for creating the national forests: Clearcutting was blamed for floods, and publicly employed "scientific foresters" were expected to use superior management techniques.[1]

Except in the Douglas-fir region, the Forest Service predominantly used selection cutting techniques through the 1940s.[2] After about 1945, the use of clearcutting expanded, and the practice was followed by controversy wherever it went. Clearcutting was first heavily used on the west side of the Oregon and Washington Cascade Mountains, where debates over the procedure may have stimulated the growth of the wilderness movement in the 1950s.

In the 1950s, large-scale clearcutting began in the Rocky Mountains and the Intermountain West, leading in the 1960s to particularly bitter controversies in Montana and Wyoming, including the Bitterroot Forest debate briefly described in chapter 2. In the 1960s, the Forest Service switched from selection cutting to clearcutting in the eastern hardwood forests. Particularly intense controversies in West Virginia led to a successful lawsuit against the practice in 1974. Congress passed the National Forest Management Act in 1976 to make clearcutting legal again. During the 1970s, selection cutting in California was gradually replaced by clearcutting. This change has caused controversies that are only now reaching a crescendo.

Wherever the Forest Service has applied clearcutting, then, it has faced intense political opposition. The fact that clearcutting often requires herbicides to ensure prompt reforestation causes even greater controversy. Yet the agency has steadfastly defended the practice and has refused — against the advice of its attorneys — to seriously consider alternatives.

The most commonly cited alternative to clearcutting is selection cutting. Under this system, individual trees or very small patches of trees — under 5 acres and preferably under 1 acre — are selected for cutting. At any given time, trees of many different ages can be found in a selection-cut forest. For this reason, selection cutting is often called *uneven-aged management.* Conversely, since all the trees grown following a clearcut are about the same age, clearcutting is called *even-aged management.* Other methods of even-aged management include *shelterwood cutting*, which consists of removing about two-thirds of the trees in one harvest. The remaining trees shelter seedlings from temperature extremes and provide a seed source for natural regeneration. After seedlings are established — five to ten years after the first cut — the shelter trees are removed.

The agency's preference for clearcutting has spawned a mythology that is completely refuted by its own research. According to a 1974 Forest Service pamphlet titled *Patience and Patchcuts*, for example, Douglas-fir "has evolved with fire and other natural processes that periodically create openings flooded with sunlight. Douglas-fir needs such openings in which to reproduce." Patchcuts — clearcuts averaging 40 acres in size — "approximate nature's clearcuts."[3]

The reality, according to Jerry Franklin, is very different. Franklin is the chief plant ecologist at the Pacific Northwest Forest and Range Experiment Station, a part of the Forest Service's research branch. Writing with Dean DeBell, then a researcher with Crown Zellerbach and more recently with the Experiment Station, Franklin says that "biologically, no types or species appear to require large clearcuttings for successful regeneration — by 'large,' we mean clearcuttings that exceed ten acres."[4]

David Smith, a forestry professor at Yale University and author of the standard forestry textbook on silviculture, says, "It is actually fortunate that the routine of clearcutting, burning, and seeding or planting of Douglas-fir has worked at all. In most instances the optimum environment for young Douglas-firs is found underneath partial shade."[5]

If this statement is true, then shelterwood cutting, not clearcutting, is closer to the optimum system for Douglas-fir. The large fires described in *Patience and Patchcuts* rarely destroyed every tree in a forest. Instead, they left many living trees and many large snags (dead trees), both of which provided shelter for young seedlings. Smith points out that "if silviculture were a perfect imitation of natural processes leading to the ecological optimum for each species and site, a number of variants of the shelterwood method rather than clearcutting would be the most common kind of silvicultural management of the [Douglas-fir] region."[6]

Franklin goes so far as to say that selection cutting may be suitable for Douglas-fir in some instances. "There are even xeric [dry] habitats within the heart of the Douglas-fir region where selective management of Douglas-fir is possible and (perhaps) even essential."[7]

Originally, concern over the ecologically best harvest methods stemmed from a desire to minimize reforestation costs. Some trees, such as western hemlock, are considered *shade-tolerant*, meaning that seedlings will readily sprout and grow in heavy shade. Other trees, such as red alder, are *shade-intolerant*, with seedlings requiring full sunlight to grow. Douglas-fir is intermediate in tolerance because it can stand more shade than alder but has a disadvantage when competing in shade with western hemlock. Research has found that Douglas-fir seedlings grow best under about one-third shade until they are ten years old, then they grow best in full sunlight — precisely the conditions provided by a shelterwood harvest.

Shade tolerance may actually be a measure of seedlings' ability to compete for water, nutrients, and other factors as well as sunlight. Yet the shade tolerance of species provides a good guide to the *natural succession*, or progression of the forest from one species or group of species to another. In many Douglas-fir forests of the Oregon and Washington coast regions, for example, an opening such as a clearcut is often quickly occupied by red alder. After 50 to 100 years, the alder, which cannot regenerate beneath itself, might be replaced by Douglas-fir. The long-lived Douglas-fir may dominate the site for 500 to 800 years and then be replaced by hemlock. Since hemlock can regenerate in dense shade, it becomes the final stage, or *climax* forest — until another disturbance such as a serious fire or clearcut. Douglas-fir may respond well to clearcutting when red alder and other shade-intolerant species are not abundant. As Franklin points out, Douglas-fir can grow well under a selection system on dry sites that are unfavorable to hemlock. On many other sites, shelterwood cutting may produce regeneration most rapidly.

Despite this variation, the Forest Service has applied clearcuttings almost exclusively to Douglas-fir forests. As a result, many productive forests have been converted to fields of alder, salmonberry, tan oak, and other species whose commercial value is low or zero. To prevent this process, the Forest Service has established a policy of hand planting of seedlings after almost every clearcut. As Smith notes, "The widespread use of planting actually obscures the fact that complete exposure is actually detrimental to regeneration of Douglas-fir on many sites."[8] Such plantings are frequently accompanied by applications of herbicides to prevent alder and other noncommercial species from occupying the sites.

Clearcutting of Douglas-fir may be least justified in the interior valleys of southwest Oregon, where low rainfall prevents hemlock from successfully competing with Douglas-fir yet also creates ideal growing conditions for tan oak, manzanita, and other noncommercial species. Here the Forest Service plans to apply herbicides to each clearcut twice — once before and once immediately after planting. Exposed soil temperatures rise to over 150 degrees — lethal to seedlings — so the agency places shade cards over the seedlings to replicate shelterwood conditions.

Forest Service officials often claim that clearcutting is the most efficient practice. It is true that sale preparation and logging costs are lower. However, reforestation costs after clearcutting can be much higher than after other harvest systems, such as shelterwood cutting. Reforestation of clearcuts in southwest Oregon easily costs over $500 per acre and can reach $1,000 per acre.[9] Yet the trees that will grow are unlikely ever to be worth that much when their value is discounted to the present at 4 percent.

Considering these problems, the Forest Service's heavy reliance on clearcutting might be puzzling. But it is easily explained if the Forest Service is a budget maximizer: By using clearcutting, the budget-maximizing manager has two opportunities to augment the budget. First, clearcutting allows the manager to stretch sale preparation and administration funds, which are appropriated by Congress, to as many acres as possible. Preparing a clearcut for sale may cost only $10 per thousand board feet, while preparing a shelterwood cut may cost $20 per thousand board feet and a selection cut even more. Preparing a clearcut for sale is less expensive than other cutting systems because only the perimeter of stands to be cut needs to be marked rather than individual trees. More volume is removed per acre, so fewer acres need to be traversed or monitored to arrange and administer a given sale volume.

Second, each additional acre that is sold makes the largest possible contribution to the K-V fund, since reforestation is so costly. If shelterwood cutting were used on certain forest types, stands could naturally regenerate a large percentage of the time. With clearcutting, the chance that brush may occupy the site requires immediate large investments in reforestation.

Thus, given limited funds from Congress, more acres can be arranged for sale using clearcutting. Given high per acre K-V deposits, clearcutting provides the greatest possible contribution to the Forest Service budget.

Forest managers who propose clearcuttings may not consciously do so to maximize their budgets. Over the years, however, those managers who use clearcutting may tend to be more successful within the agency, in part because clearcutting increases agency budgets. This success may have affirmed their belief in the practice and inspired others to support clearcutting as well.

Wilderness

Although the Forest Service is often characterized as antiwilderness, the agency showed that it was willing to experiment with the idea of managing for wilderness characteristics in the 1930s and 1940s. Arthur Carhart, a Forest Service official, is credited with the idea of proposing that a lake that was being contemplated as the site of a developed resort would be more valuable if it were managed as a scenic wilderness.[10] The first established wilderness is attributed to Aldo Leopold, who, as supervisor of the Gila National Forest, created the Gila Wilderness area.

In the 1930s, Bob Marshall became the Forest Service's director of recreation and began proposing wilderness areas throughout the National Forest System. Marshall, who was also a founder of the Wilderness Society, encountered only mild opposition to the idea. The timber industry was in the midst of a depression, and there was little demand for national forest timber. By 1939, when Marshall died of a heart attack at age 38, the Forest Service was managing 15.9 million acres of land as wilderness, primitive areas, or similar designations.[11]

According to Marshall's biographer, James Glover, wilderness was attractive to the Forest Service of the 1930s for at least two reasons. First, outdoor recreation was becoming popular, and groups such as the Sierra Club, the Appalachian Mountain Club, and the Izaac Walton League formed at least as powerful a constituency as the timber industry. Second, wilderness was seen as an alternative to transferring lands out of Forest Service control to the Park Service. The Park Service advocated making national parks out of parts of national forests in the Olympic Mountains of Washington, the boundary waters area of Minnesota, the White Mountains of New Hampshire, and the Sierra Nevada of California. Some of these proposals succeeded, but others did not — at least in part because the Forest Service dedicated lands to wilderness.[12]

After the war, however, the rising demand for national forest timber led Forest Service managers to remove heavily timbered areas from the wilderness system. In the mid-1950s, the Willamette National Forest removed the French Pete drainage from the Three Sisters Wilderness. Local recreationists, who were used to hiking in that area and did not want to see it logged, strongly protested, but the Forest Service insisted that timber management was the best use. This and similar controversies led to passage of the 1964 Wilderness Act, which provided legislative protection to many forested wilderness areas. Although they have been controversial, many further additions to the wilderness system have been made since that time.

In 1969, Congress passed the National Environmental Protection Act (NEPA), which required all federal agencies to prepare "a detailed statement" describing the environmental impacts of "major" actions that would have a "significant" effect on the environment.[13] In 1972, a federal court ruled that building roads in roadless areas qualified as "major" and "significant" actions, and that the Forest Service would have to evaluate roadless areas for wilderness before developing them.[14] The Forest Service has made several attempts to comply with this ruling, including two nationwide roadless area reviews and evaluations (RARE, which led to the 1972 court ruling, and RARE II), a land-use planning process during the 1970s, and forest planning during the 1980s. Court rulings have declared the RARE processes inadequate under NEPA, and other rulings have suggested that the land-use plans are also inadequate.

One example of the Forest Service's attitude toward wilderness and roadless areas can be found in a series of memos between the Region 6 office and the Olympic National Forest at about this period of time. The Shelton Ranger District of the Olympic Forest was part of the Shelton Cooperative Sustained Yield Unit, and Simpson Timber Company considered all the timber in the district to be part of its private reserve. When the time came to prepare a plan for the Shelton Unit, the Olympic Forest asked the region if the roadless areas could be ignored.

The region replied that the court cases were clear: All roadless areas had to be considered for wilderness. But the Olympic wrote another memo warning that Simpson Timber Company would probably sue if Shelton roadless areas were even considered for wilderness. Although the region again responded that roadless areas must be evaluated for wilderness suitability, the Shelton Plan never mentioned the existence of the roadless areas. New road construction soon eliminated the roadless areas, and environmentalists did not learn about the roadless areas until after the roads were built.[15]

Forest Service evaluations of wilderness values were strictly qualitative. But John Krutilla, of Resources for the Future, and Anthony Fisher, of the University of Maryland, developed a method of quantifying wilderness recreation values.[16] Krutilla and Fisher pointed out that the decision to develop a potential wilderness area was irreversible, and so the decision should take into consideration the value of future roadless recreation. To do so, Krutilla and Fisher estimated the rate at which roadless recreation use was growing. The value of all future recreation use could be discounted back to the present and compared with the present value of commodity resources whose development would conflict with the recreation values.

Krutilla and Fisher recognized that recreation opportunities that required roads might be available after the area had been developed, but in the cases they studied, the supply of available developed recreation facilities far exceeded the demand. Thus, adding more developed recreation capacity would produce no increase in recreation values.

The demand for the commodities that could be extracted from roadless areas might also be growing, but the decision not to develop an area is reversible, whereas the decision to develop an area is irreversible. Moreover, consumers are not concerned with whether the timber or minerals they consume come from a roadless area or an area that is already developed, whereas roadless recreationists might place a high value on the unique qualities of a particular roadless area.

For these reasons, Krutilla and Fisher saw no reason to project the future demand for commodities. Instead, they compared the present value of commodities at current prices with the present value of amenities, considering future demand for the roadless area. When the present value of future demand for roadless recreation exceeds the present value of timber at current prices, the area should not be developed, leaving open the option of future recreation use. This option should be closed if, at some point in the future, development values exceed the present value of future recreation.

Krutilla and Fisher applied their methods to proposed dams and mineral developments but not to timber. When the Forest Service proposed in 1977 to prepare a new roadless area review and evaluation — RARE II — CHEC suggested that Krutilla and Fisher's methods be used. However, the Forest Service again decided to use strictly qualitative measures of wilderness.

To show that a quantitative analysis is possible, CHEC evaluated each of the 177 roadless areas in Oregon national forests. Forest Service estimates of sustained yield capacities, timber values, fire management costs, recreation use, and other values for each roadless area were supplemented by interviews with national forest engineers and timber managers to obtain the best estimates of the cost of logging, the numbers of miles of road that would be required to provide access to the roadless timber, and road costs per mile. Although the Forest Service sometimes built roads to high standards to allow recreation use after the timber had been removed, CHEC specified that estimated road costs should reflect only the road standards needed to remove the timber. Based on this information — including the fact that the Forest Service estimated that roads would increase fire management costs in nearly all of

Figure 10.1 Recreation & Timber Values in Oregon Roadless Areas

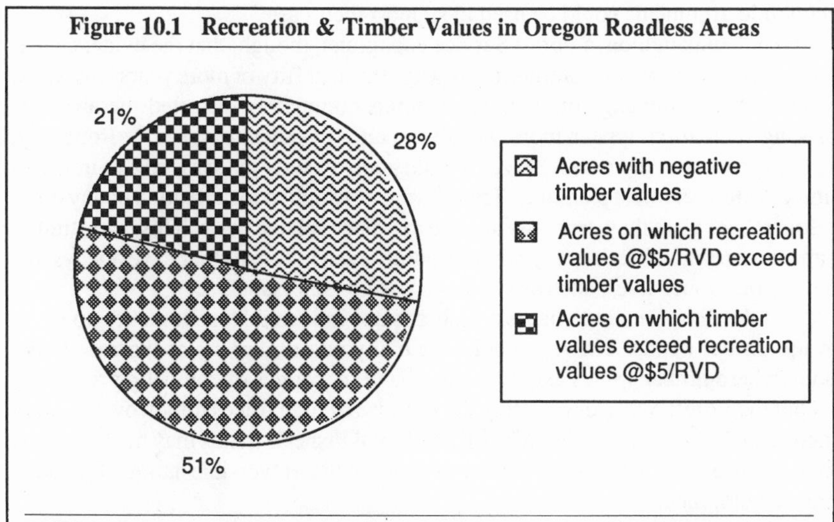

Timber values are are exceeded by primitive recreation values in over half of Oregon's roadless areas if recreation is assumed to be worth $5 per visitor day. This figure is based on the relatively high timber values of the late 1970s. Recreation values would stand out even more when compared with today's lower timber prices.
Source: Randal O'Toole, *An Economic View of RARE II* (Eugene, OR: CHEC, 1978), p. 27.

the roadless areas — CHEC concluded that road construction and timber sales would lose money in nearly a third of Oregon's roadless acres.[17]

That was only the first step, however. Many of the roadless areas were already visited by numerous recreationists — one had an estimated 88,000 visitor days of use each year. Moreover, some of Oregon's designated wilderness areas, particularly in the high Cascades, were receiving near-capacity use. While many people suggested that the most productive national forestlands should be managed for timber and the least productive lands should be reserved for recreation, the fact was that Oregon's population was located near the most productive lands — which therefore also had the highest recreation values.

Using Krutilla and Fisher's formulas, CHEC estimated the future recreation use of each roadless area in Oregon. Assuming that primitive recreationists were willing to pay about $5 a visitor day, the present value, discounted at 6 percent, of all future recreation use was calculated, assuming that the area was dedicated to primitive use. This was compared with the present value of all future timber harvesting, assuming that the area was dedicated to timber.

The calculations found that 2.4 million acres of Oregon roadless lands, out of a total of 3 million, appeared to be more valuable for recreation than they were for timber (figure 10.1). Oregon wilderness advocates had always been told that timber was much more valuable than recreation, and their desires would have to be sacrificed to meet the nation's growing needs for wood. Although environmentalists were aware that the Forest Service lost money on timber it sold in Colorado and Alaska, this was the first time Oregon environmentalists realized that economic arguments could be used to support their case.

Kent Connaughton, a Forest Service economist, thought that the evidence that recreation values would continue to grow for the next fifty or more years was weak. According to Connaughton, recreation trends completely distorted the analysis, making recreation appear more valuable than any other resource.[18] Ironically, several years later the Forest Service would apply exactly the same sort of trends to timber values in forest planning. Trends were not applied to recreation or any other resource, nor were they generally applied to timber-related costs. This made timber appear more valuable than any other resource, even in forests that have always lost money on timber sales, such as Colorado's Pike–San Isabel.

Yet the decision not to harvest timber is not irreversible, so it makes no sense to use timber trends. Harvests could begin anytime timber prices increase to the point where timber values exceed costs — and the values of competing resources — but the Forest Service was using the promise of high prices tomorrow to justify increased timber sales today. While Krutilla and Fisher's trends may have been too high, the use of trends was appropriate because of the irreversible nature of roadless area development.

A federal judge in California ruled that RARE II was inadequate under NEPA.[19] But Congress passed wilderness bills for many states that declared RARE II "sufficient," allowing the Forest Service to begin to develop roadless areas in those areas that had not been designated wilderness. The Forest Service's accelerated road construction policy, described in chapter 8, may have been aimed partly at removing the future option of designating many roadless areas as wilderness. This was certainly alleged by many critics in the early 1980s, after wilderness bills with "sufficiency" language had been passed for many states. Although such legislation typically specified that such areas should be reevaluated for wilderness in ten years if they were still roadless, the Forest Service seemed intent on constructing roads in many of them before that ten-year period ended.

The various Forest Service regions took very different attitudes toward roadless areas in forest planning. Although the wilderness bills had released the Forest Service from the obligation of reviewing roadless areas for possible wilderness designation, this did not mean that the areas were considered by Congress to be suited only for timber. Environmental groups believed the Forest Service should consider backcountry or other designations in at least some alternatives of the forest plan environmental impact statements. Region 6 agreed and directed forests to consider roadless areas for either "primitive" or "semiprimitive" recreation.[20] The primitive designation would prohibit roads or timber cutting under any circumstance until the plan was revised. The semiprimitive designation would ordinarily not allow roads or timber cutting, but exceptions could be made in emergencies, such as insect epidemics or fires.

Attitudes in most other regions were very different. Regions 3, 8, and 9, in particular, acted as though all roadless areas released by Congress should be managed for timber without any alternatives being considered. The term *roadless areas* could not even be found in many plans for these regions.

The Chequamegon Forest (WI) Plan assumed that all roadless areas would be managed for "multiple-use." The EIS noted that planners assumed that 4.25 miles of road per square mile would be needed for multiple-use management based on "a 1982 analysis of logging road densities."[21] Clearly, to Chequamegon planners, "multiple use" means roading and logging. Similar assumptions were made for Region 9 forests. In fact, planners for the Nicolet Forest (WI) hotly denied that any roadless areas existed in their forest even though 19,000 roadless acres had been inventoried in RARE II.[22]

In North Carolina, the Nantahala and Pisgah forests contained less than 115,000 acres of wilderness, research natural areas, and other undeveloped areas when the draft plan was published. No additional acres were proposed for roadless management in any of the alternatives. Instead, all acres were apparently open for timber cutting.[23] Although this was changed by the final plan, publication of that plan was delayed for nearly two years because of opposition from the Reagan administration.

Over 750,000 acres of New Mexico's Gila National Forest were considered roadless in RARE II, yet the term *roadless area* can hardly be found in the draft Gila Forest Plan or EIS. A large share of the roadless acres were forested, and the Gila Plan proposes to manage 98 percent of the forested acres for timber. Thus, without any consideration for alternative uses of those roadless acres, the Gila effectively was deciding to develop all areas.[24]

These questions might seem less important if the timber in the roadless areas had high values. But the Gila, Nantahala, Pisgah, Nicolet, and Chequamegon forests typically spend far more on timber management than they collect in sale receipts. Even without considering wilderness values, taxpayers would be better off if the roadless areas were left undeveloped. But, as noted in chapter 4, overestimated timber prices and other planning assumptions biased most plans against protection of roadless lands.

In sum, despite the fact that the concept of designated wilderness was practically invented by the Forest Service, the agency developed a strong antagonism to the concept in the 1950s and 1960s. This attitude expressed itself in the 1970s and early 1980s in an unwillingness to seriously consider forested roadless lands for any use but timber management — even when the timber values on these roadless lands were negative.

Both the Forest Service's early support for and later antagonism toward wilderness can easily be explained if the Forest Service is a budget-maximizing agency. When timber was unimportant but the agency's jurisdiction was threatened by the Park Service, support for wilderness was an important defense of the land base and, indirectly, of its budget. Later, when timber became more important, enthusiasm for wilderness diminished because wilderness makes a much smaller contribution to the Forest Service budget than timber. While the agency might support adding nontimbered acres to wilderness, each forested acre would make a much larger contribution to the agency budget if it were managed for timber than if it were managed for wilderness.

Budget maximization may not have been the motive of any particular individual in the Forest Service who made decisions about wilderness and roadless areas. But a growing wilderness system appeared to protect and possibly increase the agency's budget in the 1930s, while budgets in the 1950s could be increased by reducing the number of acres dedicated to wilderness. Thus, while each individual may have had a variety of motives, the agency as a whole always followed the course that would maximize its budget.

Grazing

After timber, grazing is the leading commodity use of the national forest in terms of acres managed. Under the Public Rangelands Improvement Act (PRIA) of 1978, grazing fees are determined by a formula that is similar to the Forest Service's residual pricing system in that a producer's cost index is subtracted from a beef price index. The resulting fee has been $1.35 per animal unit month (AUM) for the past several years. Although PRIA expired in 1986, President Reagan issued an executive order that effectively established the $1.35 figure until Congress sets another price.

Grazing is particularly controversial in Arizona and New Mexico, where charges of overgrazing are supported by forest planners. According to planning documents, current grazing use on nine of the eleven forests in those two states exceeds grazing capacity by an average of 25 percent (figure 10.2). An extreme case is the Tonto National Forest (AZ), whose grazing use has been 65 percent greater than capacity. "Capacity" is defined by the Forest Service as "the maximum stocking rate possible without inducing damage to the vegetation or related

Figure 10.2 Overgrazing in New Mexico and Arizona Forests

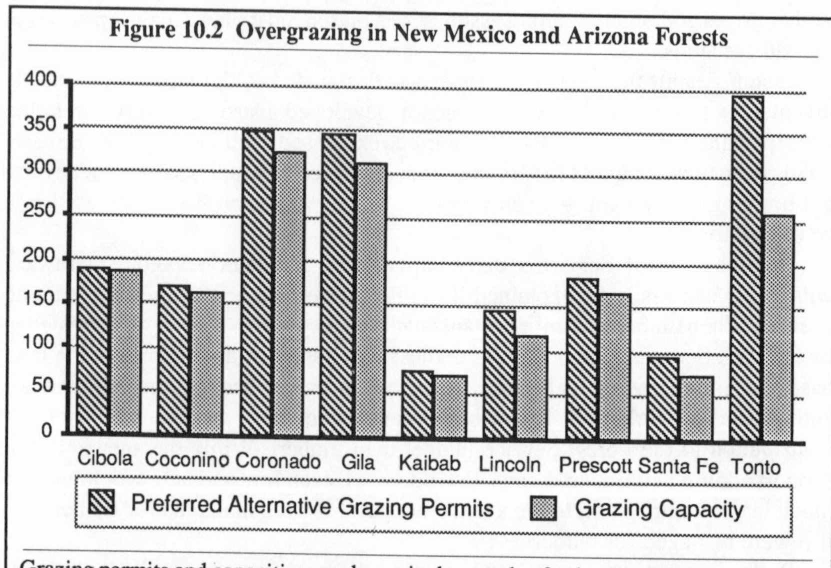

Grazing permits and capacities are shown in thousands of animal unit months per year. Although most forest plans in Region 3 admit to overgrazing, the preferred alternatives for those plans fail to reduce grazing use to capacity levels.
Source: Forest Plan EISs for indicated forests.

resources."[25] Yet many of the draft plans prepared for these forests refused to consider any alternatives that would reduce grazing use to capacity levels.

Most planners agreed that reductions in grazing use to capacity would have unacceptable negative effects on local communities. Cibola National Forest planners, for example, refused to consider an alternative that would reduce grazing use "because of serious social and economic impacts on the livestock permittees and communities dependent upon the ranching industry."[26] However, planning documents revealed that the effects of grazing on local communities was negligible: For example, only three jobs depend on Cibola overgrazing.

Planners also insisted that demand for grazing already exceeded current use. Coconino National Forest (AZ) planners, for example, claimed that "the projected level of grazing is less than grazing demand because of funding limitations."[27] As with timber, demand is being expressed as a single point, when actually the quantity of demand for grazing would decrease if grazing fees were increased to fair market value. No forest plan, of course, ever contemplated raising fees to fair market value.

Rather than reduce grazing use, planners proposed increased budgets that would be used for improvement to raise rangeland capacity to current use levels. Large amounts of money would be needed to accomplish this: The Tonto Plan calls for reducing grazing use by 7 percent and spending $540,000 per year (in 1987 dollars) to increase capacity — an average of $3.90 for each overcapacity AUM, or almost three times grazing receipts.[28] Such expenditures may not be made if Congress does not fund the 20 to 50 percent increases in total budgets required to implement most forest plans, but even if the expenditures are not made, grazing will continue at proposed levels.

All of these actions — underpricing of the resource, claims of high demand, reliance on local community stability (and therefore on appeals to Congress to protect constituents), and requests for higher budgets — are symptomatic of budget-maximizing behavior. As with timber, the Forest Service has two budgetary incentives to maintain or increase grazing. First, range administration costs are funded by a Congress that is sensitive to the lobbying power of ranchers eager to maintain below-cost grazing. Second, by law half of all grazing collections are retained by the forest manager for range improvements.

Since grazing fees are only $1.35 per AUM, the fund for range improvements is not large. In fact, Congress set a floor on this fund of $10 million per year for all Bureau of Land Management and Forest Service range improvements, a floor that has been greater than half of all range collections for several years. But individual managers know that their share of the $10 million depends on the number of AUMs grazed in their forests and districts each year. A reduction in grazing means a direct reduction in range improvement funds, giving managers an incentive to propose increases in appropriations rather than grazing reductions to make capacity equal to use.

Recreation

Forest Service research indicates that recreation is the most valuable resource in most national forests. Yet forest plans rarely give recreation the attention it deserves. Managers appear to consider recreation as little more than a side-effect of timber management, even though this is clearly untrue.

Six national forests in the northern Rocky Mountains have particularly high recreation values because of their proximity to Yellowstone National Park. According to forest planners, Yellowstone forest recreation produces four times the benefits of timber and grazing combined (figure 10.3).[29] Recreation also provides sixteen times as many jobs as timber (figure 5.4). Despite these impressive figures, Yellowstone forest planners propose that four times as much money be spent on timber as on recreation (figure 10.4).[30] This strategy puts timber management in the red and reduces recreation opportunities, yet the proposal is rational from the view of budget maximization.

The Land and Water Conservation Act authorizes the Forest Service to collect recreation fees only for developed recreation.[31] For example, fees may not be collected in national forest campgrounds unless toilets, garbage collection, and other facilities are provided. The act also provides that collected fees go into the Land and Water Conservation Fund to be used for recreation facilities.

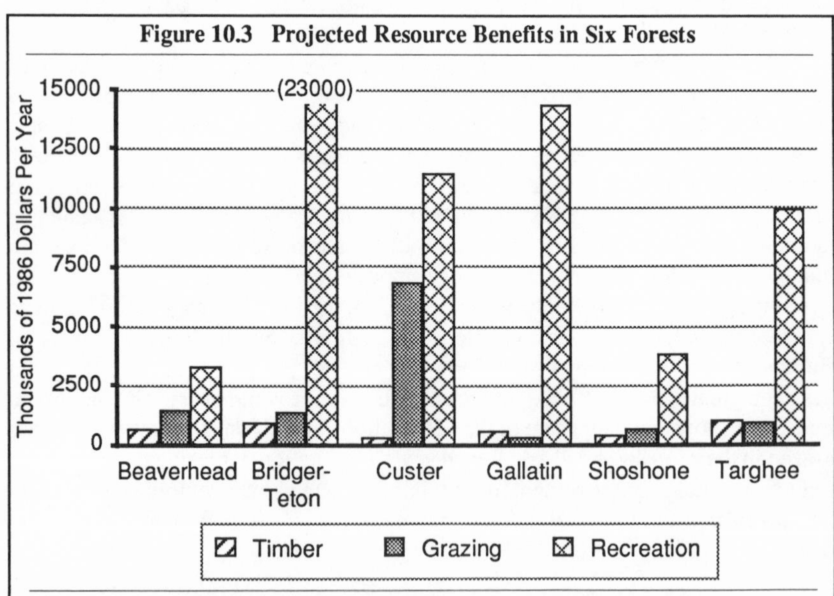

Figure 10.3 Projected Resource Benefits in Six Forests

Recreation benefits are projected by forest planners to be four times the total benefits from timber and grazing in the six forests near Yellowstone National Park.

Source: CHEC, *Economic Database for the Greater Yellowstone Forests* (Eugene, OR: CHEC, 1987), p. 21.

Figure 10.4 Planned Resource Budgets in Six Forests

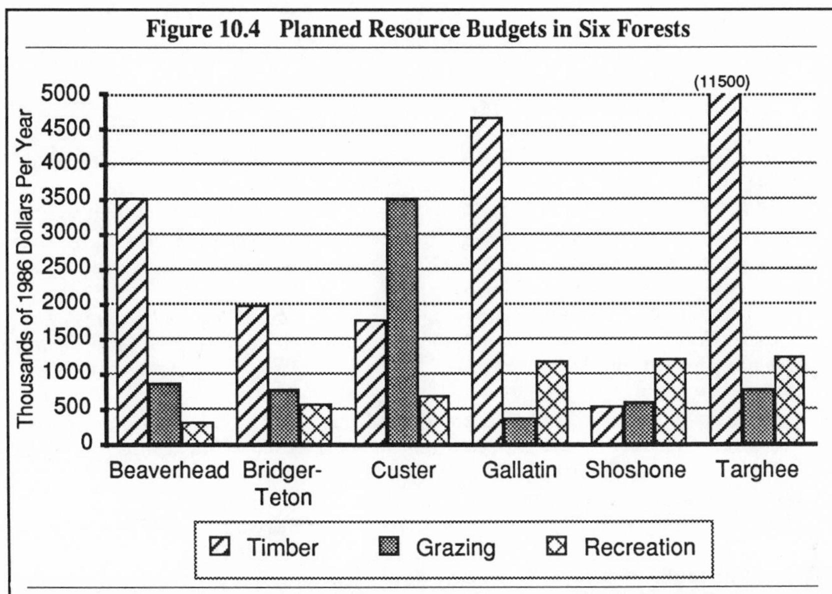

Timber and grazing budgets average four times recreation budgets in the Yellowstone forests. Note that in most cases timber management budgets are greater than the timber benefits shown in figure 10.3.

Source: CHEC, *Economic Database for the Greater Yellowstone Forests* (Eugene, OR: CHEC, 1987), p. 22.

In practice, forest managers see no tie between fees collected and appropriations from the Land and Water Conservation Fund. Expenditures from the fund are determined politically rather than by the managers who collect the fees. While timber managers can augment their budgets by selling trees and range managers can augment their budgets by grazing cattle, recreation managers have no hope of supplementing their budgets by attracting recreationists. Since this is the case, forest supervisors and district rangers are likely to consider recreation a burden rather than an opportunity.

Summary

Budget maximization logically explains Forest Service policies and attitudes toward clearcutting, wilderness, grazing, and recreation. Timber primacy seems to conflict with the nondeclining flow policy, while the idea that the Forest Service is ruled by a "conservation ethic" conflicts with Forest Service range and recreation programs and proposals. Although the evidence is circumstantial, budget maximization appears to be confirmed by the fact that no other theory explains all of these policies.

Notes

1. Ashley L. Schiff, *Fire and Water* (Cambridge, MA: Harvard University Press, 1962), p. 120.
2. David Clary, *Timber and the Forest Service* (Lawrence, KS: University Press of Kansas, 1986), p. 181.
3. USDA Forest Service, *Patience and Patchcuts* (San Francisco, CA: Forest Service, 1974), pp. 3–5.
4. Jerry F. Franklin and Dean S. DeBell, "Effects of Various Harvesting Methods on Forest Regeneration," *in Proceedings from a Symposium on Even-Age Management*, ed. Richard K. Hermann and Denis P. Lavender (Corvallis, OR: Oregon State University School of Forestry, 1973), p. 35.
5. David M. Smith, "Applied Ecology and the New Forest," *in Proceedings of the Western Forest and Conservation Association, Western Reforestation Coordinating Committee*, 1971, pp. 3–7.
6. Ibid.
7. Jerry F. Franklin, "Effects of Uneven-Aged Management on Species Composition," *in Uneven-Aged Silviculture and Management in the Western United States* (Portland, OR: USDA Forest Service, 1977), p. 67.
8. Smith, "Applied Ecology," pp. 3–7.
9. USDA Forest Service, *Rogue River National Timber Resource Final Draft Environmental Impact Statement* (Corvallis, OR: Forest Service, 1978), p. J-2.
10. John V. Krutilla, *Resource Availability, Environmental Constraints, and the Education of a Forester* (Washington, DC: Resources for the Future, 1977), pp. 27–28.
11. James M. Glover, *A Wilderness Original: The Life of Bob Marshall* (Seattle, WA: The Mountaineers Books, 1986), p. 265.
12. Ibid., p. 94.
13. National Environmental Policy Act of 1969, 42 U.S.C. 4332.
14. *Sierra Club* v. *Butz*, 3 Environmental Law Reporter (Environmental Law Institute) 20,071 (N.D. Cal. 1972).
15. "Shelton Ranger District," unpublished memo from Regional Forester to Olympic Forest Supervisor, 12 April 1976, 5 pp; "Shelton Ranger District Land Use Plan," unpublished memo from Olympic Forest Supervisor to Regional Forester, 19 May 1976, 2 pp; "Shelton Ranger District Land Use Plan, "Regional Forester to Olympic Forest Supervisor, 22 July 1976, 5 pp.
16. Anthony C. Fisher and John V. Krutilla, "Valuing Long Run Ecological Consequences and Irreversibilities," *Journal of Environmental Economics and Management* I:96–108.
17. Randal O'Toole, *An Economic View of RARE II* (Eugene, OR: CHEC, 1978), p. 27.
18. Kent Connaughton, presentation at Western Forest Economists meeting in Wemme, OR, 1979.
19. *California* v. *Bergland* 483 F. Supp. 465 (E.D. Cal. 1980) aff'd *sub nom.* CA v. Block 690 F. 2d 753 (9th Cir. 1982).
20. "Regional Planning Direction," memo from Regional Forester to Region 6 Forest Supervisors, 10 November 1983, p. 26.
21. USDA Forest Service, *Chequamegon National Forest Plan Draft Environmental Impact Statement* (Park Falls, WI: Forest Service, 1985), pp. B-22, C-1.
22. Interview with Terry Moore, Nicolet Forest Planner, October, 1985.
23. CHEC, "Review of the Nantahala-Pisgah National Forests Draft Plan and EIS" (Eugene, OR: CHEC, 1984), p. 14.
24. CHEC, "Review of the Draft Gila Forest Plan and EIS" (Eugene, OR: CHEC, 1985), p. 13.
25. USDA Forest Service, *Coconino Forest Plan Final Environmental Impact Statement*

(Flagstaff, AZ: Forest Service, 1986), p. 100.
26. USDA Forest Service, *Cibola Forest Plan Final Environmental Impact Statement* (Albuquerque, NM: Forest Service, 1986), p. 126.
27. USDA Forest Service, *Coconino Forest Plan* (Flagstaff, AZ: Forest Service, 1986), p. 18.
28. USDA Forest Service, *Tonto Forest Plan Final Environmental Impact Statement* (Phoenix, AZ: Forest Service, 1986), pp. 145–47.
29. CHEC, *Economic Database for the Greater Yellowstone Forests* (Eugene, OR: CHEC, 1987), p. 21.
30. Ibid., p. 22.
31. Land and Water Conservation Act of 1965, section 4(b) (16 USC 460*l*-6a).

Part III

Prescribing the Cure

Chapter Eleven
An Irrational Process

In 1981, forest planners and economists were shocked to hear Richard Behan — then dean of the School of Forestry at Northern Arizona State University — propose the repeal of the National Forest Management Act (NFMA). The planning process, said Behan, "cannot be made to work and cannot be patched up." This is not what planners, many of them recently hired by the Forest Service, expected to hear.

Young and energetic, the planners and economists were full of optimism about how they could improve forest management. But their enthusiasm was dampened by Behan's speech, delivered at a meeting of forest economists near Portland, Oregon. Behan warned that NFMA contained a perfect model of a planning process. "But a corollary of statutory perfection has become apparent lately, and it is sobering indeed: An imperfect plan is an illegal plan, and if there is a fir tree in Oregon that will mean litigation later."[1]

When Congress passed NFMA in 1976, it contemplated an objective planning process that would solve conflicts over wilderness, clearcutting, herbicides, below-cost sales, and other national forest issues — issues Congress hoped it would not have to settle itself. Heavy public involvement in the plans would ensure that national forest management would be in the public interest.

More than ten years later, the results fall far short of expectations. Although the process was designed to be objective, most plans appear to slant data and analyses in favor of commodities. Plans cost at least three times as much as the Forest Service first estimated and require many more years to complete than expected. Effective public participation has been circumvented by the technical complexity of the plans. Major issues, such as clearcutting and wilderness, have been sidestepped or ignored.

These problems are not due to inept staff or corrupt bureaucrats. Instead, the planning process itself is an impossible dream. The process promotes polarization, ignores economic realities, and effectively cheats taxpayers out of a fair rate of return on their investments in the national forests. Moreover, although the Forest Service is promising greater efficiency in management, in fact it is incapable of fulfilling such promises.

The Rationality of Planning

As originally conceived, forest planning appears to be an entirely rational process. Forest Service planning rules issued in 1979 describe an elegant, ten-step procedure, including data gathering, assessment of resource supply and demand, formulation of alternatives, and publication of draft and final plans for public comment. As noted by Forest Service economist Dennis Schweitzer, writing with economist Hanna Cortner, "Traditional planning theory espouses the ideal of value-free technicians making decisions based on comprehensive and objective assessments of problems." Yet this is unrealistic: "Because planning allocates benefits to some and imposes costs on others, planning is inherently political."[2]

Cortner and Merton Richards add that rational planning is flawed because it "imposes crippling data demands" and "becomes obsessed with techniques" — a fair description of the Forest Service process, which typically spends $150,000 on computer time just to prepare one plan.[3] Behan emphasized this point by calling planning "a procedural black hole which threatens to suck in all the money, manpower, and data which comes near."[4]

The Forest Service, of course, takes advantage of this "black hole." Just as NFMA gave Congress a way out of making hard decisions on essentially political issues, so FORPLAN gives the agency an escape from the same decisions. Debates about clearcutting, herbicides, wilderness, and grazing are all subsumed in a morass of computer printouts generated by a "black box" computer program that most people — including most Forest Service officials and planners — cannot understand.

Cortner and Richards warn that, while rational planning is supposed to be "zero-based," actual decisions tend to be incremental. People who "believe that a rational planning process can bring about significant shifts in program emphasis are likely to be disappointed when confronted with incremental results."[5] But black box computer models allowed the Forest Service to tell what it hoped would be an unwary public that the "plan is cost-efficient" because FORPLAN said so. As one planner described the process, "Our motto is, 'Garbage In, Gospel Out.' "[6]

For example, many planning teams believed that time was too short to gather forest inventory data prior to completing the plan. Instead, they relied on outdated timber inventory data that may have been filled with errors and, being over ten years old, was subject to many changes because of recent cutting and growth. Ironically, delays in the plan often resulted in the plan's completion *after* a new inventory was done.

The Beaverhead Forest expected to finish its forest plan in December 1981. The final plan was not actually released until May 1986, just a few months before its new timber inventory was completed. The Santa Fe Forest once hoped to have its final plan complete in March 1981. In reality it was not completed until early 1987, about the same time as its timber inventory. The timber inventories used in these plans were obsolete, while the new inventories were essentially ignored by planners.

New inventories are important because they often reveal major errors in the previous database. The Deschutes Forest (OR) does not expect to publish a final plan until late 1987. A new timber inventory has been completed that shows dramatic changes from the data used in the plan —but it will probably not be incorporated into the final plan. Plans for the Grand Mesa-Uncompahgre-Gunnison, San Juan, and many other forests were also prepared without the benefit of a recent timber inventory.[7]

Assessments of demand are just as unrealistic. Although the planning rules specify that planners should estimate demand for various resources as a "price-quantity relationship,"[8] planners were subsequently directed by the Washington office to assume a "horizontal" timber demand curve.[9] This means that no matter how much timber is sold from a forest, planners assume that timber companies will pay the same price. Naturally, the demand for all other resources is assumed to be horizontal as well.

Environmental impact statements for plans frequently fail to fulfill legal requirements that they consider a "broad range of alternatives." Although grazing is a major use of many national forests, most plans assume that permitted grazing use will be nearly constant in all alternatives. This is especially surprising in plans for Arizona and New Mexico forests, many of which planners estimate are being grazed at 10 to 25 percent above their sustained yield capacity (see figure 10.2).

Although the planning process is designed to be rational, in practice it ignores many proper planning procedures. In too many cases, plans resemble the economic analyses whose rules are changed so that the results "are, in their managers' judgment, more realistic."[10]

The High Cost of Planning

The Forest Service originally estimated that forest planning would cost about $100 million, or about $1 million per plan. It now appears that many plans will cost well over $3 million, and total planning expenses probably approach $300 million. A 1986 study by the General Accounting Office found that Idaho's Boise and Clearwater national forests each spent about $2.5 million on planning between 1981 and 1985. Both forests began planning prior to 1981 and neither had completed their plans by 1985 — or, for that matter, by 1987 — so total planning costs may possibly exceed $4 million each. While plans in Regions 2, 8, and 9 probably cost less because they were completed sooner, plans in Regions 5 and 6 may cost much more.

The expense of planning is often not justified by the resources that are being planned. The average forest plan may cost close to $3 million, which is more than most national forests collect in timber and other receipts. Since plans are to be revised every ten years, many forests are spending more than 10 percent of their annual income on planning — certainly far more than is justified by the results.

There is some indication that many plans may require significant amendments or revisions long before they have been in effect for ten years. Direction for plan implementation proposed by the Forest Service says that "sustained differences between proposed and actual budgets" will require a "significant amendment." Such an amendment is less than a full revision but requires planners to undertake all ten steps of the planning process, including preparation of alternatives and publication of draft and final environmental impact statements.

Most forest plans are proposing budgets that are 20 to 50 percent greater than historic levels in real dollar terms.[11] Considering the 1985 Deficit Control Act and

Congress's current general desire to reduce deficits, such increases are extremely unlikely. Significant amendments will be needed to take these budget shortfalls into account, and so the entire planning cycle could begin again just after it is completed.

Time Enough to Do It Twice

According to Robert Wolf, who helped write NFMA as a staff member with the Congressional Research Service, when NFMA was written "a lot of us thought that the plans would all be underway by now and some in their eighth or ninth year."[12] Instead, the first plans were not completed until 1983, and some may still be unfinished in 1990.

In 1980 California Forest Service officials confidently predicted that the Shasta-Trinity Forest Plan would be completed in October 1981. Its first draft was rejected by the Washington office in 1982, however, and planners were forced to completely revise it. A later draft reached the public in late 1986, but pressure from the timber industry led to its withdrawal. A new plan being prepared will not be completed before — and possibly well after — March 1989.

Other plans may require even longer to complete. Most plans for Oregon and Washington forests were to be completed by 1983. As this is written in 1987, no Region 6 plan is near completion. Controversies over proposed reductions in timber sale levels have led to repeated timber industry attempts to delay these plans, and few planners are optimistic enough to believe that many plans will be completed before 1989.

Most planning delays have been political. In 1982 the Reagan administration completely revised the regulations, written when Carter was president, that planners were to follow. The new rules were largely written by Doug MacCleery, the former timber industry lobbyist who had been made deputy assistant secretary of agriculture in charge of the Forest Service.

Not satisfied with the Forest Service's progress, MacCleery wrote a memo (over John Crowell's signature) to the chief in 1983 berating the Forest Service for poor-quality plans. "Appropriate changes will be made," said the memo, "unless review drafts begin to improve significantly soon."[13] Many in the agency read this as a veiled threat to replace the chief with a political appointee. The Forest Service responded to the memo by trashing many plans that were ready to go to the printer and starting over from scratch. Noting that the first plans were written under strict deadlines, planners complained that "there's never enough time to do it right, but always enough time to do it twice."[14]

Individual plans were felled when MacCleery or other USDA staff members — often tipped off by local industry officials concerned about timber harvest levels — tested their axes against them. Planners for the Mt. Hood, Shasta-Trinity, Boise, and numerous other forests considered their efforts wasted when told to do their work over again. Many forests lost valuable expertise when entire planning staffs quit or transferred in frustration. Although they cynically spoke of NFMA as "the forest planners full employment act of 1976," many planners were pessimistic about their future careers in the Forest Service.

Public Nonparticipation

A central feature of the planning process and the method to be used to resolve conflicts was an extensive system of public participation. In practice, forest plans are so complex that real public participation is all but impossible. A typical forest plan, along with its environmental impact statement and appendices, is three to four inches thick. The plans often exceed a thousand pages in length, mostly solid text unrelieved by photos or other graphics. Depending on their predilection, readers have a choice of pouring over hundreds of pages of boring prose or attempting to assimilate scores of pages of numbers, which are often printed in tiny type. Experts at readability reviewed the 1985 draft Santa Fe Forest Plan and concluded that reading it required a nineteenth grade education.[15]

As if the plans are not formidable enough, most public questions about the plans are answered with references to the FORPLAN computer model. A typical FORPLAN run is a two- to three-inch thick bundle of eleven-inch by fifteen-inch computer paper, filled largely with cryptic codes and numbers that have no meaning to a lay reader. Most forests produce between thirty and a hundred such computer runs. Requests from the public for the data that went into the model are often filled by providing another computer printout consisting entirely of numbers with little explanation.

Only a few FORPLAN experts exist outside the Forest Service. Most work for the timber industry, and no more than three work for environmental groups. The number of citizens' groups who have been able to penetrate FORPLAN's mysteries without the help of outside consultants can be counted on the fingers of one hand. And FORPLAN is not the only computer model used in planning. HYSED, IMPLAN, RMYLD, ECOSIM, Prognosis, RAM-PREP, DF-SIM, and a myriad of other programs defy anyone to attempt analysis.

Ironically, computer models may actually make review easier for a dedicated member of the public. When not relying on computer models, many forest planners use "professional judgment," which defies analysis or explanation. Models, at least, can reveal the assumptions made by planners when making decisions. But for most members of the public, FORPLAN and other models are simply not worth the time needed to understand them. As a result, public participation in most plans devolves into a contest to see which interest group can generate the most form letters. As many as 80 percent of the comments on many plans are form letters or form postcards.

Unresolved Issues

The issues of wilderness, clearcutting, herbicides, and below-cost sales have been public concerns since well before NFMA was passed. Yet, as has been suggested elsewhere in this book, these issues have been all but ignored in most forest plans.

Wilderness — or, to be precise, the disposition of the hundreds of thousands of acres of national forest roadless lands — has been one of the central controversies of national forest management since the mid-1950s. The fact that it is more expensive to build roads into roadless areas than to harvest timber from areas with roads should be a central element of timber management planning. Yet numerous forest plans have virtually ignored the fact that some lands in the national forests remain without roads. Chapter 10 pointed to the Gila (NM) and Chequamegon (WI) plans as examples. In many other cases, such as the draft Nantahala-Pisgah Plan, roadless areas were acknowledged in planning documents, but the cost of road construction was not included in planning computer models.

Clearcutting, too, has been ignored. The National Forest Management Act was passed in response to debates over clearcutting. Yet many forest plans barely acknowledged the alternatives to clearcutting. In January 1985 the USDA Office of General Counsel (OGC) sent a letter to the chief of the Forest Service noting that "we recall no forest plan which has included an alternative which uses selective cutting as its principal harvesting method." Yet, said the OGC, "a substantial segment of the public seems to oppose even-aged management as a predominant harvest method."[16] The letter suggested that uneven-aged management alternatives should be a part of future forest plans. In the two years since that time, only one forest plan has included such an alternative.

Herbicides are a major issue because of evidence that the chemicals used are carcinogenic, mutagenic (mutation-causing), or teratogenic (birth-defect-causing). Since the national forests provide watersheds for many municipalities and many more small domestic water supplies, these concerns are real and often expressed with strong emotions. Much herbicide spraying is a direct result of the decision to use clearcutting as a harvest tool. A report prepared by Region 5 (California) once estimated that substitution of shelterwood and selection cutting for clearcutting would reduce herbicide needs in that Region by 80 percent.[17]

Although forest plans are deciding (usually without considering alternatives) to use clearcutting, few plans are considering the effects of or alternatives to herbicides. Instead, herbicide planning takes place separately in so-called *vegetation management plans*. These plans, prepared after the clearcuts are in place, are limited to considering various alternative methods of brush control. The alternative of using shelterwood cutting or other methods cannot be seriously considered since the clearcuts have already been made.

Planners for the Tahoe National Forest (CA) prepared timber yield tables and other data necessary to consider a no-herbicide alternative. However, they were directed by the regional office to exclude this information from the plan.[18] Similarly, officials in Region 6 (Oregon and Washington) have directed planners not to consider alternatives with and without herbicides.[19]

Below-cost sales have also been ignored by many forest plans. Chapter 4 told how the San Juan (CO) Plan claimed positive net benefits from high levels of timber cutting even though FORPLAN computer runs indicated the benefits would be negative. Many other forest plans covered up negative cash flows by using price trends and high timber prices.

The Ozark-St. Francis (AR) EIS, for example, failed to list timber benefits and costs by decade but promised that overall present net benefits from the timber program would be positive. Yet the forests' computer model indicated that timber would produce a negative cash flow during the next ten years. Use of the 1980 RPA price trends, however, led planners to conclude that future benefits would more than pay for present-day losses.[20] This was not disclosed by the EIS.

In sum, four of the most important issues facing national forest management were not only unresolved by forest plans, they were not even addressed. Plans also fail to address such issues as limited budgets, range management, and water quality. The benefits of planning are certainly questionable when important controversies such as these are not settled or even considered.

The Forest Activist's Dilemma

An important reason why planning cannot resolve conflicts is that it is in the interest of all the major participants for issues to be polarized. Agency officials want to have as much flexibility as possible, and polarization effectively increases the decision space within which they can appear to be reasonable or moderate. Interest groups who seek compromise are castigated by their peers, while those who promote polarization find their members most responsive when told to "donate money or your [job] [wilderness] [ranch] [favorite recreation area] will die."

Even those interest groups that are willing to cooperate find it difficult because political questions are pervaded with "prisoner's dilemmas." In game theory, the prisoner's dilemma is a situation in which two people have a choice of cooperating or not cooperating. Although neither is allowed to know in advance which the other chooses, the reward from cooperating or dissenting depends on the choice the other makes.

Suppose, for example, that timber sales and wilderness withdrawals on a particular national forest could *both* be increased by 20 percent — a situation that is not uncommon. In fact, as described in chapter 4, timber sales and nontimber acres could both be increased in any forest whose plan falls inside — rather than on — that forest's production possibilities frontier; this includes most forests in Regions 1 and 8 and several in Regions 3 and 5. In such situations, the Forest Service will often be persuaded to change its plans only if encouraged to do so by both environmentalists and industry. Congressional lobbying may also be required to designate wilderness or to increase timber funding. Since environmental and industry lobbyists are fairly evenly balanced, the cooperation of both is needed to realize timber and wilderness increases.

But timber industry leaders know that if wilderness areas are created with their help, environmentalists could later make further gains by reneging on agreements and lobbying Congress for reduced timber funding. In the eyes of the environmentalists, this would improve the environmental quality of the forests. The environmentalists also know that if they lobby Congress for more timber funding, industry leaders could quietly subvert efforts to create new wilderness.

Thus, both sides are wary of working with the other. But if they don't work together, they are likely to have no more wilderness and no more timber. The result is the decision matrix shown in table 11.1. Note that, if both sides cooperate,

Table 11.1 The Forest Activist's Dilemma

		Timber Industry	
		Cooperate	Dissent
Environmentalists	Cooperate	20% more wilderness 20% more timber	20% more wilderness 10% less timber
	Dissent	0% more wilderness 30% more timber	0% more wilderness 0% more timber

On many forests, it is still possible to increase both timber sale levels and wilderness acreage. But such increases are not politically feasible unless environmental and industry groups work together. However, industry suspects that environmentalists will cooperate only until the wilderness is designated, and then betray the industry to try to reduce timber sales. Environmentalists suspect that industry will feign cooperation, but will secretly work for increases in timber sale levels at the expense of wilderness. Thus, neither side cooperates and the result is neither side gets as much of the timber or wilderness as would be possible.

wilderness acres and timber sales will each be increased by 20 percent. But given that one side either will or will not cooperate, the outcome for the other side is best if it does not cooperate. Thus, neither side will cooperate, and there will be no increase in timber or wilderness.

Prisoner's dilemmas are encountered repeatedly in interest group politics, and interest groups almost invariably refuse to cooperate. Even though intensive management of high-site timberlands could produce more timber and leave more land for nontimber purposes than extensive management of a greater number of acres, environmental and industry groups fail to work together to achieve that goal.

Prisoner's dilemmas are also encountered in daily business transactions. For example, a store could sell groceries for a fixed sum, leaving both the store and the customer wealthier because they have made a mutually beneficial exchange. But the store might be tempted to sell spoiled food, which is available at a lower cost, and refuse to replace it when customers complained. At the same time, its customers could write bad checks to pay for the food. If the pattern followed the political system, all transactions would consist of spoiled food being exchanged for bad checks. Yet such refusals to cooperate almost never happen. Because they are interested in generating goodwill, many stores go out of their way to ensure that customers are satisfied. Customers who expect to buy from the store again do not deliberately write bad checks.

Why do prisoner's dilemmas lead to cooperation in the business world but not in the political world? Robert Axelrod is a political scientist who has extensively studied the prisoner's dilemma. He notes that noncooperation pays off in the short run, but cooperation pays if transactions are repeated an indefinite number of times.[21] The repetition gives the business actors an incentive to cooperate so that they will not receive spoiled goods or bad checks on the next transaction. Many environmental battles are fought on a "last stand" basis—this is the last chance to save a particular roadless area or this is where industry must draw the line to stop reductions in sale levels — so there is no perceived opportunity for repeated transactions.

Axelrod notes that the probability that a transaction will be repeated is functionally equivalent to a discount rate.[22] A discount rate of 10 percent represents a probability that the transaction will be repeated ten times (1/0.1) — or, if the rate is 10 percent per year, for ten years. A 10 percent discount rate thus represents a ten-year planning horizon, while a 5 percent rate represents a twenty-year planning horizon. This horizon is much longer than the political horizon, which is effectively two years because the political balance in Congress changes every two years. Thus, political opponents in a local environmental battle may never look past the one or two transactions they will make in the next two years, after which all the rules will change.

Politics does not always lead to noncooperation. Members of Congress conduct several hundred transactions with one another each year, and these repetitions provide many opportunities for cooperative vote trading. Compromises made by environmental lobbying groups in Washington, DC, are a reflection of this opportunity for cooperation. At a national forest level, however, there is little room for such cooperation since forest plans, for example, are prepared only once every ten years.

Polarization may be especially tantalizing to interest groups because it leaves open the possibility of a big victory. Although a 20 percent increase in wilderness and timber may be the most efficient way of managing a national forest, there is always the chance that timber or wilderness could be increased by 30 percent. Although such an increase would be inefficient — perhaps requiring below-cost sales or producing a surplus of costly wilderness — the interest groups promoting the increase would not have to pay the costs. Instead, taxpayers in general would pay the cost in lost receipts or added expenditures. Thus, such inefficiencies represent a transfer of wealth from the general taxpayer to specific interest groups.

Such wealth transfers can be found throughout the National Forest System. Below-cost timber sales, below-cost grazing, and below-cost recreation cost taxpayers hundreds of millions of dollars each year. Yet the cost to individual taxpayers is small — perhaps $10 to $20 per year — so taxpayers have little incentive to stop this transfer.

Entrepreneurship in the Forest Service

The Forest Service has developed a new program of "entrepreneurship" that it promises will increase the efficiency of national forest management. The program was conceived in 1985 when top Forest Service officials, concerned that the agency was losing its reputation as an "excellent" organization, decided to adopt some of the recommendations of Waterman and Peters's book *In Search of Excellence*. The resulting entrepreneurship program was first tested on the Gallatin (MT), Mark Twain (MO), and Ochoco (OR) national forests.

Under the program, forest officials are not required to spend appropriations according to the strict line item budget approved by Congress. Instead, managers are told to achieve targets as efficiently as possible. Any funds left over after the targets are met can be spent on any activities the managers choose. If funds remain at the end of the year, the managers can carry them over to the next year.

These changes can improve management efficiency. Since managers no longer need to adhere to line item budgets, several staff positions on each forest were freed from red tape to work on other projects. Since funds can carry over from one year to the next, managers no longer have the incentive to spend their entire budget each year just to show that they need more money the next.

At the same time, the entrepreneurship program is too little too late. Managers are still required to meet targets, and the appropriateness of the targets themselves is never questioned. Forest Service Associate Chief George Leonard admits that "our best opportunity to improve efficiency is in our internal management, not in resource management. We're stuck with policies and procedures for dealing with our resources that have evolved over eighty years, and realistically they give a tremendous inertia to our system."[23]

In addition, no attempt is made to change the existing incentives provided by the Knutson-Vandenberg Act and other laws. Managers who wish to increase their budgets will discover that they can use their appropriations to sell timber and collect more K-V funds. In fact, the system creates a new incentive because most of the targets are commodity-oriented, so managers faced with budget cuts are likely to transfer money intended for amenity resources to timber and other commodities.

The Forest Service Cannot Reform Itself

The entrepreneurship program may be a move in the right direction, but by itself it could lead to even worse abuses than are found today. Forest management will remain inefficient, promoting artificial resource shortages, damaging the environment, and transferring wealth from taxpayers to special interests — often to the relatively rich. Reform of the Forest Service will require congressional legislation that creates new incentives for proper forest management rather than depends on the benevolence of managers and the cooperation of interest groups.

In 1981 Behan's judgment that NFMA should be repealed appeared premature. In 1987 it appears prophetic. Forest planning is a costly, time-consuming process that is failing to solve any of the problems with national forest management, while people who spend enormous energies on planning find that all their efforts have moved the Forest Service less than a fraction of a percent. If anything, planning is one more indication that the Forest Service is a budget-maximizing institution.

Chapter 12 discusses some of the reforms that have been proposed in the past and the philosophies behind those ideas. Chapter 13 presents a new concept of reform that blends economics with environmental goals of diversity and protection for all species. Chapter 14 turns this concept into a concrete series of ten reforms, nine of which could be implemented in a single congressional act. Chapter 15 projects the effects of these reforms on the economy and national forest environments.

Notes

1. Richard Behan, "RPA/NFMA — Time to Punt," *Journal of Forestry* 79:802–805.
2. Hanna Cortner and Dennis Schweitzer, "Institutional Limits and Legal Implications of Quantitative Models in Forest Planning," *Environmental Law* 13:493–516.
3. Hanna Cortner and Merton Richards, "The Political Component in National Forest Planning," *Journal of Soil and Water Conservation* 38 (2):79–81.
4. Richard Behan, "Time to Punt," p. 803.
5. Hanna Cortner and Merton Richards, "Political Component," p. 80.
6. Interview with William Connelly, Umpqua National Forest Operations Analyst, March 1984.
7. Randal O'Toole, "Forest Service Uses Ancient Timber Inventories," *Forest Watch* 6(11):5.
8. 36 *Code of Federal Regulations* 219.12(e)(3).
9. *Forest Service Manual* 1971.65.
10. Clark Row, "Forest Service Budget Maximization: A Dissent," *Renewable Resources Journal*, 2(4):5–7.
11. CHEC, *Economic Database for the Greater Yellowstone Forests* (Eugene, OR: CHEC, 1987), p. 27.
12. Interview with Robert Wolf, 3 August 1987.
13. "National Forest Land Management Planning," unpublished memo from John Crowell to the Chief, 19 January 1983, p. 7.
14. *Forest Planning* staff, "Does the Forest Service Still Have a Mission?" *Forest Planning* 5(1):13-16.
15. Anonymous, "Testimony Criticizes Santa Fe Plan," *Forest Watch* 6(10):3.
16. "Observations on the Legal Sufficiency of Forest Plan EISs," memo from the USDA Office of General Counsel to the Chief of the Forest Service, 25 January 1985, 4 pp.
17. USDA Forest Service, *Forest Reestablishment on National Forests in California* (San Francisco, CA: Forest Service, 1974), p. 61.
18. CHEC, "Review of Tahoe National Forest Planning Data" (Eugene, OR: CHEC, 1985), p. 14.
19. "Treatment of the Herbicide Issue in Forest Planning," memo from Region 6 to forest supervisors, 2 August 1984, 2 pp.
20. CHEC, "Economic Analysis of the Ozark-St. Francis Forests Plan" (Eugene, OR: CHEC, 1985), p. 1.
21. Robert Axelrod, *The Evolution of Cooperation* (New York, NY: Basic Books, 1984), p. 59.
22. Ibid., p. 42.
23. Randal O'Toole, "Interview With the New Chief and Associate Chief of the Forest Service," *Forest Watch* 7(11):15–20.

Chapter Twelve
Challenging the Dominant Paradigm

Environmentalists who are concerned about national forest problems have traditionally proposed to solve them through prescriptive legislation. Such proposals have rarely received wide support, in part because Congress has preferred to leave technical issues to its experts — in this case, Forest Service professionals. In addition, laws that regulate practices rather than incentives treat the symptoms and not the causes of forest problems and thus may be doomed to fail.

The 1980s have seen two new groups whose means and objectives appear to be quite different from each other as well as from more traditional environmental groups. The New Resource Economists, as represented by the Political Economy Research Center of Bozeman, Montana, work primarily in the academic world and may be best known for their proposals to sell the national forests. Earth First!, a major faction of the Deep Ecologists, primarily take direct actions against environmental destruction, such as sitting in front of bulldozers, and strongly prefer public ownership. Although these two groups appear to represent polar extremes, in fact there are many similarities between them.

These three views appear to present significantly different alternatives for reforming the Forest Service. Individually, none of them is feasible. However, it may be possible to blend the best elements of each view into a proposal that is politically viable, economically efficient, and environmentally sound.

The Failure of Prescriptive Legislation

Popular support for prescriptive legislation is probably due to the belief that Forest Service leaders are timber primacists who need restrictions to prevent them from taking economically or environmentally unacceptable actions. A number of prescriptive proposals were in the bill sponsored by West Virginia Senator Jennings Randolph during the debates over the National Forest Management Act in 1976.[1] Largely written by environmental groups, this bill was strenuously opposed by the timber industry and the Forest Service as being too "prescriptive." The final legislation, which focused on a forest planning process, included few prescriptions. Yet it is unlikely that the Randolph bill would have corrected any of the problems with national forest management.

Perhaps the most prescriptive section of the Randolph bill would have prohibited clearcutting in the eastern mixed hardwood forests. Yet the bill merely required detailed environmental assessments prior to clearcutting elsewhere. The final legislation struck the clearcutting proscription but retained the requirement for environmental assessments and other clearcutting limits.

The Randolph bill would have required that trees, or most trees in a stand, be "dead, mature, or large" before cutting. "Mature" was defined as a tree that has reached the peak of its vigor and is declining, and "large" was defined as trees that are the size of those in a natural mature stand. The final legislation required that trees not be cut before the stands they are in reached "culmination of mean annual increment" — their peak of growth. The difference between the two is that some individual trees may peak much later than entire stands. In this case, the final legislation was just as prescriptive as the Randolph bill, but the prescription is slightly different.

The Randolph bill would have required the Forest Service to limit timber sales to the nondeclining yield levels of each ranger district. The final legislation limited timber sales to the nondeclining yield levels of each national forest. Again, the two are equally prescriptive, but the prescriptions are slightly different.

The Randolph bill would have required that all trees be marked before cutting. The final legislation allows the Forest Service to mark only the perimeters of clearcuts before cutting. While the Randolph bill is more prescriptive, the additional prescription existed only to make clearcutting less economically efficient, and therefore less attractive, than other forms of cutting.

The Randolph bill prohibited "type conversion" of eastern hardwood forests to pine and required that other proposed type conversions receive detailed analyses and public review. The final legislation struck the proscription but retained the analysis and review requirements.

The Randolph bill required the preparation of forest plans, as did the final legislation, and prohibited higher levels of the Forest Service or USDA from establishing targets for the national forests. This was struck from the final legislation, and many people have accused the Forest Service of using RPA objectives as targets. However, many national forest plans that failed to meet these objectives have been approved.

Finally, the Randolph bill required the Forest Service to develop a cost accounting system to compare the benefits and costs of timber sales. It also required the Forest Service to report below-cost timber sales to Congress each year. The final legislation eliminated the cost accounting system but retained the reporting requirement. In 1984 Congress appropriated $400,000 for the Forest Service to develop a cost accounting system. As described in chapter 2, the system that was developed was strongly biased toward timber.

The Randolph bill illustrates two major problems with prescriptive legislation. First, either it is so specific that its writers must be experts on the management of every national forest or it is so vague that it is meaningless. For example, in proscribing clearcutting from eastern hardwood forests, the Randolph bill was proclaiming Congress to be a better expert on forest management than the Forest Service. Yet by requiring just an interdisciplinary review prior to clearcutting elsewhere, the Randolph bill was creating a vague standard that could be enforced only by requiring bureaucrats to create mountains of paperwork.

Second, by trying to change forest practices without changing incentives, prescriptive legislation treats symptoms rather than causes. Recognizing that the Forest Service often sells timber at a loss, the Randolph bill would have required the Forest Service to develop a cost accounting system. But the *Timber Sale Program Information Reporting System*, described in chapter 2, shows that managers with an incentive to sell below-cost timber can be trusted to produce a cost accounting system that obscures such losses.

This is demonstrated by numerous cases where prescriptive legislation was passed by Congress but ignored by the Forest Service. For example, a good case can be made that below-cost sales are already illegal. Cross-subsidized sales also appear to violate the *Forest Service Manual.* Yet such sales take place routinely in every Forest Service region. Section 14(a) of the National Forest Management Act requires that timber be sold for not less than "appraised value." As described in chapter 8, legislative history and *Forest Service Manual* directives indicate that appraised value means fair market value, the price at which a buyer and seller would agree to exchange. Since no reasonable seller would agree to sell at a loss, below-cost sales appear to violate this provision.

The *Forest Service Manual* also specifically and directly proscribes cross-subsidized timber sales. Section 2403.25 of the manual states that "the appraised value of a tract of timber will not be reduced to obtain utilization of a species, size, or class." This is called the *tract value policy*. Section 2422.56(3) of the manual gives, as a specific example, a sale that includes 5 million board feet of pine, which is appraised at $22 per thousand board feet, and 5 million board feet of fir, which is appraised at minus $4 per thousand.

In a typical cross-subsidized sale, if the base rates for the two species were $9 per thousand, the value of the pine would be reduced to $9 to raise the value of the fir to $9. The total sale would then sell for $90,000. But the manual says that the sale cannot be sold for less than $110,000, which is the value of the pine alone. The manual permits adjustments if selling the pine alone would reduce the value of the pine. For example, brush disposal costs might be the same for the pine alone as for the pine together with the fir. This adjustment, however, would probably be less than $20,000. Thus, the cross-subsidized sale violates the tract value policy.

As indicated in chapter 8, as many as 40 percent of national forest timber sales may be cross-subsidized and thus are likely to violate this policy. A CHEC review of recent timber sales on the Sierra National Forest (CA) found seventeen sales in two years violating the policy. In eight cases, high bid prices raised the total sale price above the calculated tract value. In the remaining nine cases, the high bid prices fell short of the tract value by an average of more than $100,000 per sale.[2]

The Knutson-Vandenberg Act and other laws give managers powerful incentives to sell timber below cost and to cross-subsidize timber. Laws against below-cost sales are as unlikely to halt such sales as existing laws and provisions of the

Forest Service Manual have halted past sales below cost.

Numerous examples exist in which the Forest Service has interpreted the law to its own benefit and has subsequently found that interpretation to be illegal. A West Virginia district court and the U.S. Court of Appeals for the Fourth Circuit found the Forest Service's practice of clearcutting to be in clear violation of a law passed seventy-five years before the court decision.[3] In Oregon a 1903 law prohibiting entry into a municipal watershed by any member of the public other than a Forest Service official was violated by the Forest Service for seventeen years beginning in 1958. In that year the Forest Service began large-scale timber operations in the watershed, operations that ceased in 1975 when a court ruled them illegal.[4]

Prescriptive legislation may attempt to make the Forest Service more efficient and environmentally sensitive. However, it will greatly increase centralization — nothing is more central than Congress deciding how each acre of the national forests should be managed. Prescriptive legislation will also reduce the flexibility of the Forest Service to respond to new research and changes in the supply and demand for forest resources. Because of these problems, prescriptive legislation will fail to make the Forest Service significantly more efficient or sensitive to environmental issues than it is today.

Private Property and Public Choice

A major difference between traditional environmental groups and the New Resource Economists is the latter's focus on incentives rather than on outcomes. "In general, public-spirited actions regarding environmental quality, resource management, and other public issues have tended to be oriented rather simplistically toward desired outcomes rather than toward processes carefully designed to produce those outcomes," say Richard Stroup and John Baden, two of the leading New Resource Economists.[5] Where environmental groups seek prescriptive legislation to obtain the outcome they desire, the New Resource Economists seek to use markets as the best process for achieving that outcome.

Although many people disagree with proposals to sell the public lands, the New Resource Economists have much to say that is worthwhile. Led by Baden, Stroup, and Terry Anderson, economists and political scientists who founded the Political Economy Research Center in Bozeman, Montana, the New Resource Economists believe that the marketplace is better at allocating resources than is interest-group politics.

According to Stroup and Baden, people acquire wealth in one of two ways: by purchasing less-valuable resources and converting them to more-valuable ones and by convincing the government to transfer resources from public to private use or from one person to another. "By moving toward political allocation and away from the rule of willing consent," they say, "we have moved from a society that rewards productive activity and willing exchange to one where many of a person's best investment opportunities lie in influencing transfer activities."[6]

The marketplace is centered around the notion of private property. Property owners "have incentives to use their resources efficiently," say Stroup and Baden.

People can make money in the market by taking low-value goods and converting them to high-value goods. Thus, the market distributes resources to those who place the greatest value on those resources. "Everyone can be made better off when goods and services are moved to higher valued uses," they conclude.[7]

"The benefits of diversity, individual freedom, adaptiveness to changing conditions, the production of information, and even a certain equity derive from this market system," say Stroup and Baden.[8] The market promotes diversity because there is no single, centralized decision maker. It preserves individual freedom since those who support and wish to participate in each activity may do so on the basis of willing consent. It is equitable because those who gain the benefits pay the costs.

The government, on the other hand, is based on coercive activity. Say Stroup and Baden, "Government, with its monopoly on sanctified coercion, has the potential for being the most efficient engine ever designed for the generation of plunder."[9] According to the New Resource Economists, timber companies, ranchers, and wilderness users find it in their interests to convince Congress to force society as a whole to pay part of the costs of timber cutting, grazing, or recreation. Thus, they become "free riders" who enjoy the benefits of the national forests with everyone else paying the costs.

New Resource Economists disagree with the traditional belief that public lands are *public goods* from which everyone benefits. "It is a common misconception that every citizen benefits from his share of the public lands and the resources found thereon," say Stroup and Baden. "But public ownership by no means guarantees public benefits. Individuals make decisions regarding resource use, not groups or societies. Yet, with government control, it is not the owners who make the decisions, but politicians and bureaucrats."[10] Chapters 8, 9, and 10 have demonstrated that this is true of Forest Service activities.

Critics of the New Resource Economists say that the "perfect market" envisioned by Stroup and Baden exists only in fantasy. But the New Resource Economists respond that many of the "market failures" that make people suspicious of markets are not market failures at all but institutional failures. For the market to work, private property rights to resources must be easily transferable.

"If rights are not easily transferable, owners may, for example, have little incentive to conserve resources for which others might be willing to pay if transfer were possible," say Stroup and Baden. "Transferability ensures that a resource owner must reject all bids for the resource in order to continue ownership and use. Thus, if ownership is retained, the cost to others is made real and explicit."[11]

The market works when rights are both privately held and easily transferable because decision makers have both easy access to information through bid and asked prices and an incentive to move resources to higher-valued uses. The New Resource Economists believe that private ownership would solve most natural resource problems.

"When a natural resource is privately owned, it is often thought that the owner has only his conscience to tell him to pay attention to the desires of others," say Stroup and Baden. "Normally, however, it is the *absence* of private, transferable ownership that leads to the resource user's lack of concern for others' desires.

Private ownership holds the individual owner responsible for allocating a resource to its highest valued use, whether or not the resource is used by others.

"If the buffalo is not mine until I kill it and I cannot sell my interest in the living animal to another, I have no incentive — beyond altruism — to investigate others' interest in it. I will do with it as I wish. But if the buffalo is mine and I may sell it, I am motivated to consider others' value estimates of the animal. I will misuse the buffalo only at my economic peril."[12] Most environmental problems, such as lack of protection for wildlife, air pollution, and poor water quality, are due to the lack of transferable property rights.

Water rights are an important example. Currently, landowners in the West can have the right to use water, but ownership is still claimed by the states. Irrigators with historic water rights defend their right to use the water even if some other use is far more valuable. If water *ownership* rights were vested with various individuals, they could sell the right to pollute the water, use it for irrigation, provide it to cities, or maintain fish habitat.

Anderson compares projected water shortages with the energy crisis of the late 1970s, which he says was partly caused by government controls of market prices. "When prices were allowed to rise to market levels, energy supplies increased and demand decreased. Before this can happen with water, ownership of the resource will have to be specified and the impediments to trade removed."[13]

Some market failures still exist, but even with their imperfections, markets are often more efficient than governments, say the New Resource Economists. The national forests, for example, were created with the idea that "scientific foresters" employed by the public could objectively determine the method of management that best meets the public interest. But to assume that managers will be altruistic, say Stroup and Baden, it must be assumed "that culture can 'rewire' people so that the public interest becomes self-interest." Instead, "Property rights theorists assume that the decision maker will maximize his own utility — not that of some institution or state — in whatever situation he finds himself."[14]

Stroup and Baden point out that, "for the bureaucrat, the tax base is essentially a common pool resource ripe for exploitation." Thus, agencies make their best efforts to mine the resource before another agency gets them. "A common pool is like a soda being drawn down by several small boys, each with a straw," explain Stroup and Baden. "The 'rule of capture' is in effect. The contents of the container belong to no one boy until he 'captures' it through his straw."[15]

These problems have caused most of the environmental destruction the United States has seen in this century. "Unconstrained by the need to generate profits, bureaucrats may ignore or exaggerate the economic efficiency of the projects they administer," say Stroup and Baden. "If government agencies were required to meet standards of economic efficiency, many of their environmentally destructive practices would not occur. In the absence of such standards, the American taxpayer is, in effect, subsidizing the destruction of the environment and enhancing the welfare of bureaucrats and special interests."[16]

Many people believe that the government can do a better job of protecting the interests of future generations than the marketplace. In fact, say the New Resource

Economists, the exact reverse is true. "A government decision maker can seldom gain political support by locking resources away to benefit the unborn," say Stroup and Baden.[17] Instead, the pressure is to use the resources as rapidly as possible to benefit constituents. The long-term costs are ignored because most terms of office are short.

In contrast, if resources will be more valuable in the future than they are today, the market will encourage conservation. Private speculators make money by deferring resource use to the future. Stroup and Baden point out that speculators "save for the future by paying a market price higher than any other bidder who seeks resources for present use."[18] As noted in chapter 11, a 10 percent discount rate effectively represents a ten-year planning horizon, whereas the political planning horizon is rarely much longer than two years.

Deep Ecology: A Fundamental Transformation

At first glance, the Deep Ecologists represent the polar extreme from the New Resource Economics. Although Deep Ecology is a broad movement which has no single representative, most people who consider themselves Deep Ecologists go beyond the prescriptive legislation proposed by traditional environmentalists to demanding "a fundamental transformation of industrial and global society."[19] Deep Ecologists claim that their movement is a "religious and philosophical revolution of the first magnitude."[20] A fundamental tenet of this religion is biocentrism, the view that "all things in the biosphere have an equal right to live and blossom and to reach their own individual forms."[21]

The Deep Ecologists are, in part, responding to the endless battles being fought by more traditional environmental groups. Peter Berg likens classic environmentalism to "a hospital that consists only of an emergency room. No maternity care, no podiatric clinic, no promising therapy: just mangled trauma cases. Many of them are lost or drag on in wilting protraction, and if a few are saved there are always more than can be handled jamming through the door." He adds that "no one can doubt the moral basis of environmentalism, but the essentially defensive terms of its endless struggle mitigate against ever stopping the slaughter."[22] Deep Ecologists believe that the environmental movement is too centralized, and they prefer decentralist, local approaches to environmental problems.

Rather than getting involved in political battles, Deep Ecologists prefer to promote their general goal of biocentric equality.[23] Of course, all life must consume other living things to survive, and Deep Ecologists do not claim that every individual plant or animal has an equal right to life. Instead, they advocate preservation of genetic diversity, species, and ecosystems.

Humans, say the Deep Ecologists, "have no right to reduce this richness and diversity [of the Earth] except to satisfy *vital* needs."[24] They add that "the term 'vital need' is left deliberately vague to allow for considerable latitude in judgment."[25] Although this appears vague, it is at least clear that the 1973 Endangered Species Act is a fundamental step in the direction of Deep Ecology.

At least some Deep Ecologists realize that a mass conversion to their religion is unlikely. Even if it happened, it could not last long. Writing in a series of essays on Deep Ecology, Garrett Hardin points out that "a species composed of pure altruists is impossible."[26] If such a species existed, it would soon be replaced by a mutation that is less altruistic.

Yet many Deep Ecologists believe that a society based on their attitudes is possible and may even have once existed. The epitome of human social existence was reached, they believe, by the hunting and gathering tribes. Hunters and gatherers may have had greater social and sexual equality, less epidemic disease, and more leisure time, say Deep Ecologists.[27]

Ironically, early hunters may have been responsible for the extinction of numerous large mammals, such as the mammoth and cave bear. Social conditions in hunting tribes were probably also less idyllic than claimed by Deep Ecologists. Yet it is significant that the hunters and gatherers represent the culture that most celebrated individuals above the social group.

Where Deep Ecology resembles a religion, it borrows its tools from the hunters and gatherers. Adherents are encouraged to make vision quests, while selected individuals are treated as shamans. Vision quests and shamans disappeared from agricultural and later societies, which often place the group higher than the individual and depend on group rites rather than individual abilities — both mystical and otherwise.

The Deep Ecologists' preference for individual freedom and action conflicts with the traditional environmental tendency toward prescriptive legislation. Prescriptive legislation is possible only through the coercive power of the government, and this, in turn, is based on the philosophical belief that society's needs are paramount to the needs of the individual. Deep Ecologists reject this belief and instead call for the promotion of an ethic or morality that will suspend the need for any such coercive activities.

Although privatization and biocentrism seem to have little in common with each other, in fact there are many similarities between the New Resource Economists and the Deep Ecologists. Each group proposes to "subvert the dominant paradigm" — a paradigm being the ideas and beliefs shared by a communty of individuals. The New Resource Economists oppose the paradigm that turns to government regulation to solve environmental problems. The Deep Ecologists oppose the paradigm that uses political compromise to gain environmental protection for scarce, nonmarket resources.

In a larger sense, both groups are responding to the dominant paradigm of interest-group politics, in which environmental battles are endlessly fought but never won. The New Resource Economists and the Deep Ecologists each seek a fundamental solution that will give resource managers the incentive to protect and produce the most valuable resources. Decentralization, both groups believe, is an important part of that solution. Decentralization promotes diversity, which enables ecosystems and society to adapt to change. Decentralization also provides efficiency in resource use because it encourages innovation and the spread of innovative ideas.

The preference for decentralization is a consequence of both groups' strong belief in individual freedom. New Resource Economists see society as merely a collection of individuals and refuse to believe in the existence of any "social good" that is more than the sum of individual goods. Similarly, the Deep Ecologists favor the rights of the individual over the rights of society — with the proviso that "individual" be defined to include other species as well. Deep Ecologists, like the New Resource Economists, tend to be anarchistic.

The similarities between Deep Ecology and New Resource Economics parallel the similarities between ecology and economics. Although the language used by ecologists differs from that of economists, it frequently translates to identical concepts. Where economists discuss efficiency, decentralization, and incentives, ecologists discuss the maximum power principle, diversity, and feedback loops.[28]

The fact that these very different terms have identical meanings underscores the obvious point that *ecology* and *economics* have the same root: *oikos*, a Greek word meaning "house." Where ecology translates to "study of the house," economics is "management of the house." Although the "houses" scrutinized by ecologists and economists differ, ecological systems are really economic systems, and economic systems are really ecological systems.

Thus, New Resource Economists and Deep Ecologists use different terminology but reach many of the same conclusions: that the current system of allocating resources is flawed and that it must be replaced by one that gives individuals the incentive to manage those resources properly. Where the New Resource Economists and Deep Ecologists disagree is in their views of public lands and markets. The New Resource Economists view public land ownership as a fundamental evil that has no value in itself and that leads to environmental destruction as well as transfers of wealth from the poor to the rich. The Deep Ecologists would apply exactly the same terms to markets.

Yet markets are compatible with public land ownership. In fact, markets are the key to reforming public land management because they most closely resemble a natural ecosystem. Unlike a centrally planned economy, a successful ecosystem doesn't require an omnipotent regulator to decide how much of each good will be produced, nor does it rely on the religion or ethics of its individual members. Instead, the ecosystem uses feedback loops to be self-regulating. Through the market system, the Forest Service can be reformed in such a way that it, too, is largely self-regulating. Such reforms can achieve the goals of both the New Resource Economists and the Deep Ecologists.

Blending New Resource Economics and Deep Ecology

Successful reforms of the Forest Service must be based on a blend of New Resource Economics and Deep Ecology. Reforms must recognize that the Forest Service is run by ordinary people who are motivated by incentives. One of the most important of these incentives is the desire of the agency to increase its budget. Reforms that try to fight the bureaucracy's natural tendency to increase its budget will be doomed to failure. Instead, they should employ this tendency so that the budget becomes part of a feedback loop: As the agency maximizes its budget, it also accomplishes social, economic, and environmental goals.

Reforms should be based on two sound ecological principles: efficiency and diversity. Just as an organism must be efficient to compete in its environment, an organization such as the Forest Service must be efficient to serve the public. The current set of laws governing the Forest Service encourages inefficiency on a massive scale. Reforms should give managers the incentive to produce the highest-valued goods or services rather than producing goods or services that are below cost or that actually have negative values. Markets can provide this incentive.

Reforms should also encourage diversity since diversity will allow local managers to respond to local situations and to develop innovative techniques to improve forest management. The Forest Service achieved its reputation as an "excellent organization" largely because it was once the most decentralized agency in the federal government. Many of the current problems with the agency are due, perhaps indirectly, to increasing centralization over the last ten to twenty years. A return to decentralization can make the agency more responsive to public demand and changing tastes.

Diversity and efficiency will make the Forest Service more environmentally sensitive. Yet reforms should also recognize that some national forest resources, such as endangered plant and wildlife species, may be valuable yet difficult to sell in a marketplace. The Deep Ecology principle that all species are valuable is important here. The protection of these resources should not be neglected in reforms designed to make the agency more efficient or more diverse. Markets can make the Forest Service more sensitive than it is now to the lack of public demand for such goods as lodgepole pine and to the increasing public demand for such goods as recreation. The Endangered Species Act can protect those species not protected by markets just as well in national forests oriented to markets as in national forests oriented to interest-group politics.

Finally, reforms should recognize the irreversibility of certain forest management actions. Changes in forest policy should be designed to maintain options for the future. The reforms themselves should be reversible so that unexpected and undesirable consequences can be corrected if necessary. This means that public lands should be retained in public ownership so that public demand can easily correct any unexpected and unintended effects of reforms.

The reforms described in chapters 13 and 14 attempt to meet these standards. Markets are used to provide forest managers with feedback that will naturally lead them to manage national forests for their most valuable uses. A minimum of prescriptive legislation is required to protect endangered species and a few other truly nonmarket resources. Diversity is ensured by treating each national forest as an individual unit rather than as a part of a centrally planned organization. The resulting proposal, it is hoped, is not only economically efficient and environmentally sound but politically viable as well.

Notes

1. Senate Bill 2926, 94th Congress, 2d Session.
2. CHEC, "Review of the Sierra Forest Plan and Draft Environmental Impact Statement" (Eugene, OR: CHEC, 1986), p. 7.
3. *W. Va. Div. of Izaac Walton League* v. *Butz*, 367 F. Supp. 422 (N.D. W. Va. 1973), aff'd, 522 F. 2d 945 (4th Cir. 1975).
4. *Miller* v. *Mallery et al* ., 410 F. Supp. 1283 (D. Ore. 1976).
5. Richard L. Stroup and John A. Baden, *Natural Resources: Bureaucratic Myths and Environmental Management* (San Francisco, CA: Pacific Institute for Public Policy Research, 1983), p. 5–6.
6. Ibid., p. 3.
7. Ibid., p. 41.
8. Ibid., p. 15.
9. Ibid., pp. 2–3.
10. Ibid., p. 7.
11. Ibid., p. 17.
12. Ibid., p. 14.
13. Terry L. Anderson (ed.), *Water Rights: Scarce Resource Allocation, Bureaucracy, and the Environment* (San Francisco: Pacific Institute for Public Policy Research, 1983), p. 3.
14. Stroup and Baden, *Natural Resources*, p. 29.
15. Ibid., p. 44.
16. Ibid., p. 2.
17. Ibid., p. 24.
18. Ibid., p. 25.
19. Fritjof Capra, "The Turning Point," *Elmwood Newsletter* 1(1):1.
20. George Sessions, "Aldo Leopold and the Deep Ecology Movement," *Environmental Review*, June 1987.
21. Ibid., p. 67.
22. Bill Devall and George Sessions, *Deep Ecology — Living As If Nature Mattered* (Layton, UT: Gibbs M. Smith, 1985), p. 3.
23. Ibid., p. 67.
24. Ibid., p. 70.
25. Ibid., p. 71.
26. Garret Hardin, "Discriminating Altruism," *in Deep Ecology*, Michael Tobias, ed. (San Diego, CA: Avant Books, 1985), pp. 182–205.
27. Australopithicus, "Agriculture Led to Downfall," *Earth First!* 7(6):17.
28. Howard T. Odum and Elizabeth C. Odum, *Energy Basis for Man and Nature* (New York, NY: McGraw Hill, 1976), p. 266.

Chapter Thirteen
Marketizing the Forest Service

The Soviet Union suffers from an "excess of bureaucratic planning," says Paul Kennedy, a historian at Yale University. Planners establish output targets that agricultural collectives, factories, and other producers must meet. Because these targets are set from the top down, the Soviet economy is "unable to respond either to consumer choice or to the need to alter products to meet new demands or markets."[1]

Because workers have no monetary incentive to meet targets, worker productivity in the Soviet Union is either very poor or very inefficient. For example, although the USSR employs far more people in agriculture than the United States, per capita farm productivity is only about one-seventh that of the United States. And while Soviet industry produces huge amounts of steel, its mills consume far more energy and coal per ton of steel than those of any other country.

The USSR is also a leading producer of cement. But, as Kennedy notes, "being among the world's leading cement producers is not necessarily a good thing if excessive investment in that industry takes resources from a more needy sector." Kennedy concludes that Mikhail Gorbachev will have a difficult time reforming the Soviet economy because "the overall system suffers from concentrating on production regardless of market prices or consumer demand."

The Soviet system bears an uncanny resemblance to the Forest Service, which sets targets regardless of market prices or consumer demand. Forest managers have the incentive to make excessive investments to meet commodity targets, leading to the inefficient production of timber and forage on low-productivity lands. Meanwhile, national forest users would rather have many forests managed primarily for recreation, wildlife, and other noncommodity resources, but managers have no incentive to meet such consumer demand. Excessive investments in national forest commodities also takes resources away from more productive private lands: Although the per-acre productivity of many national forests in the Rocky Mountains is only one-seventh that of private lands in the Northwest or Southern Piedmont, the Forest Service invests more per acre in its lands than many private landowners.

Previous chapters have shown that the Forest Service reached this position through a consistent pattern: Resources are priced at less than fair market value, leading users to demand greater quantities than can be produced efficiently. With the support of user groups, the Forest Service requests increased budgets from

Congress to prevent supposed resource "shortages." These budgets allow the agency to provide resources below cost which, in turn, increases the quantities demanded and provides support for further budget increases.

This pattern is made worse by the fact that managers receive positive feedback from timber and grazing in the form of funds retained out of fees. These funds give managers an even greater incentive to sell below-cost and even negatively valued resources to increase their budgets, which, in turn, increases quantities demanded and leads managers to claim that shortages are imminent if further budget increases are not provided.

Despite the Forest Service's eighty-year history of predicting timber famines, resource shortages are the last thing the United States needs to worry about. The United States is a capital-poor, labor-poor, resource-rich country. Yet the Forest Service is squandering capital and labor in a misguided effort to "conserve" — that is, use — natural resources such as negatively valued timber and grasslands.

Peter Drucker notes that the relative scarcity of capital is often ignored by government agencies — with serious consequences for the nation. "A central economic problem of developed societies during the next twenty to thirty years is surely going to be capital formation," says Drucker. "We therefore can ill afford to have activities conducted as 'non-profit,' that is, as activities that devour capital rather than form it, if they can be organized as activities that form capital, as activities that make a profit."[2]

Underpricing of resources has insulated the Forest Service from the true demand for various forest resources. Prices and costs should act as signals to managers to tell them when they are producing too much or too little of a resource, but congressional appropriations allow managers to ignore costs, while underpricing allows managers to believe that the demand for forest resources is ever increasing.

Because the budgetary returns of timber exceed those of recreation, timber is emphasized even in those forests where recreation is far more valuable. And because the budgetary returns of a valuable timber species like ponderosa pine are nearly indistinguishable from those of a worthless species like lodgepole pine, managers do not hesitate to substitute one for the other despite the objections of purchasers.

Reforming the Forest Service requires that public demand for commodities and amenities be reflected, as perfectly as possible, in the budgets of forest managers. Ideally, budgets would increase if managers are satisfying public demands, and budgets would decline if managers are working counter to public demands.

For commodity resources, which are bought and sold in a marketplace, the best way to sensitize forest managers to public demands is by tying budgets to the net income generated from resource sales. This would give managers a disincentive to sell timber and other resources below cost. By reducing net income, any such sales would directly reduce their budgets.

For amenity resources, which are not bought and sold in a marketplace, the best way to sensitize forest managers to public demands is to create markets for those resources. The term *amenities* is traditionally used synonomously with *nonmarket resources*. In fact, environmental problems are, almost by definition, due to the failure of markets to properly allocate certain scarce resources. For this reason, environmentalists are traditionally suspicious of markets and support government actions such as regulation and resource ownership to solve environmental problems.

This chapter will show, however, that markets can be used to make national forest managers sensitive to public demands for nearly all forest resources. Endangered species represent the significant exception to this rule, and water quality may too be a problem in certain areas. Yet markets can increase protection for even these resources if only because below-cost commodity extraction will be eliminated.

The Principles of Marketizing

Because markets are the key to reforming the Forest Service, this proposal is called *marketizing*. Marketizing implies four basic changes:

1. All activities are funded out of a percent share of the net returns from user fees.
2. Forest Service appropriations from Congress are reduced to zero.
3. Managers are allowed to charge fair market value for all resources.
4. The National Forest System and other Forest Service programs are decentralized.

Funding all activities out of a percent share of net receipts will encourage managers to be efficient. For example, instead of the often expensive reforestation techniques used today, managers will use harvest methods that allow reforestation at a relatively low cost. Where current laws such as the Knutson-Vandenberg Act effectively give managers a share of gross receipts based on the number of acres harvested, this proposal would give managers a share of net receipts based on a percentage of that net.

Reducing Forest Service appropriations to zero removes the incentive to sell resources below cost to gain congressional funding. Under the current system, national forests that have no valuable timber nevertheless have a subsidized sale program so that the timber pork barrel can be spread to as many congressional districts as possible.

Allowing managers to charge fair market value for all resources will make it possible to use prices as *signals* indicating the value that members of the public place on various resources. "Fair market value" should be determined by individual managers. Because they receive a percent share of net returns, they will be motivated to keep prices high enough to cover costs and provide a return to taxpayers but low enough to sell the resource. Resources whose values are so low that no one will pay this price will not be sold.

Decentralization can be viewed as either an important goal or an indirect effect of marketization. Decentralization will allow individual national forests to experiment with new techniques of resource management. But decentralization will also be financially necessary because resource receipts will be insufficient to maintain the nearly four thousand employees now located in the Washington and regional offices.

Unlike privatization, marketization maintains the national forests in public ownership. This retains maximum flexibility and makes it possible to adjust for unforeseen circumstances. While relying primarily on markets to solve environmental problems, marketization thus leaves open the option to use the political process.

Marketization will ensure efficient production of most forest resources and an efficient allocation of forestlands. It will end inequitable transfers of wealth from the general taxpayer to special interest groups. The proposal will also end the polarization and endless battles that now characterize national forest management.

Although some people oppose the imposition of recreation fees, such fees will actually be beneficial to recreationists and will increase recreation opportunities. Efficient fee collection techniques are available, and recreation fees will swamp commodity receipts in the vast majority of national forests if the average fee collected is only $3 per visitor day. In addition, as detailed in chapter 14, marketization will protect most wildlife, wilderness, and water quality.

Marketization Is Efficient

Marketization uses economic values to determine the best distribution of resources. Purchases of national forest timber, grazing rights, or recreation permits send

Figure 13.1 Commodity and Amenity Benefits by Region

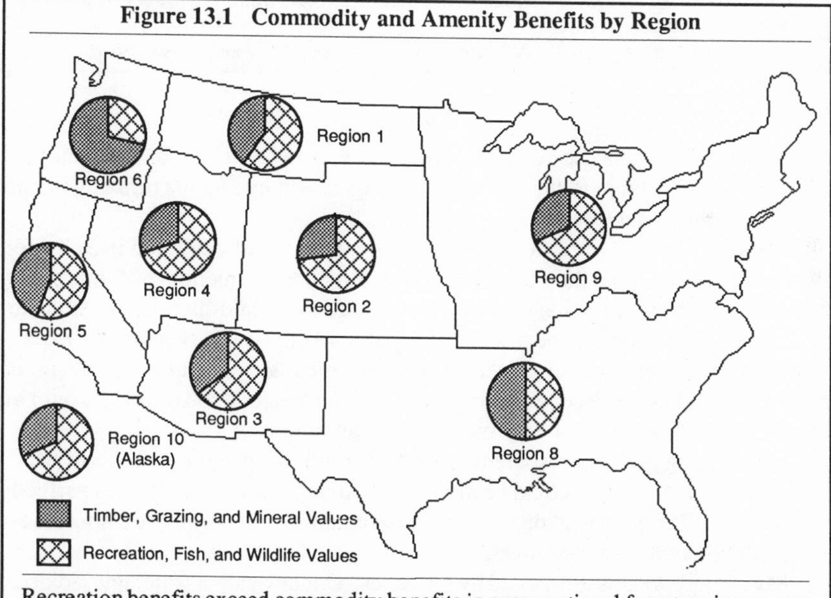

Recreation benefits exceed commodity benefits in every national forest region except Region 6. Recreation and commodity benefits are about equal in Region 8, but recreation values exceed commodity values in the northern part of the region while the reverse is true in the southern part of the region.

Source: "Total Benefits – 1985 RPA High Bound Program," unpublished document on file at Forest Service office in Washington, DC, 1987.

Figure 13.2 Returns as a Percentage of Benefits

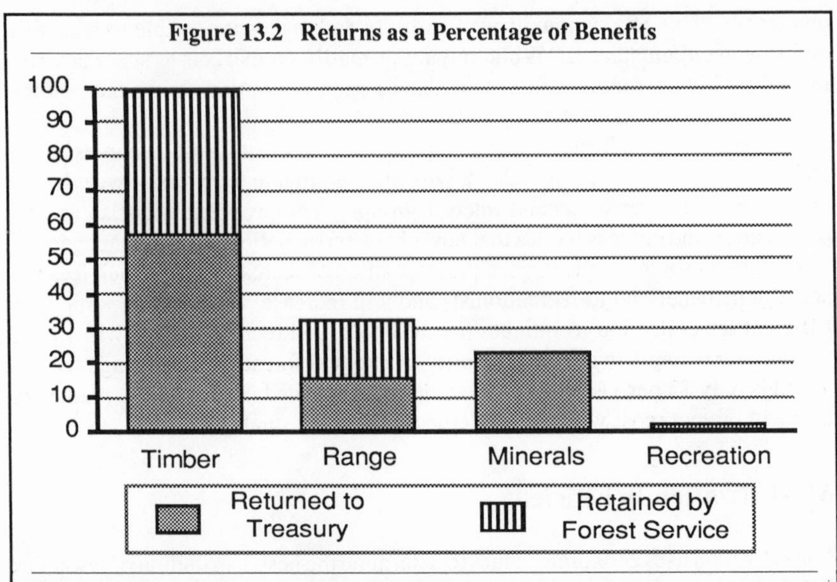

The Forest Service retains nearly half of all timber benefits and about one-sixth of all range benefits. Although it collects some fees for minerals and recreation, it retains none. Managers thus have no incentive to choose recreation over timber or grazing when the resources conflict.

Source: USDA Forest Service, *1985 RPA Program* (Washington, DC: Forest Service, 1986).

signals to forest managers, who receive a share of net receipts for their budgets. Managers who attempt to maximize their budgets will end up maximizing the net value of the national forests.

Marketization is clearly feasible for timber, which is already sold in a competitive marketplace. The only legal change would be to the method of funding timber sales and timber management. Marketization should also be possible for grazing, as private landowners sell grazing at market prices. There is some question about whether auctions of permits or other methods of setting fees would be appropriate.[3] With a decentralized Forest Service, each national forest would be free to find the best way of determining local grazing fees.

Marketizing amenity resources is more difficult to imagine. Yet most recreation, wildlife, and water could be marketized, though in some cases less perfectly than timber. To understand this, it is useful to examine just why these resources are often called nonmarket resources.

Rivers and streams are owned by the states. Under western water law, adjacent landowners may appropriate water from them, but water rights cannot be sold. Much water is thus used for irrigation, which may be its lowest value, while downstream municipalities may run short even though they are willing to pay ten to twenty times as much as the irrigators. The nontransferability of water rights also makes it impossible for fisheries interest groups to purchase minimum stream flows for fish habitat.

Similarly, fish and wildlife are owned by the states. Landowners are free to manipulate wildlife habitat, but they are often discouraged from charging fees to allow people to hunt or fish on their lands. Wyoming law, for example, forbids landowners from charging hunters, while Montana law hinders landowners from charging anglers.

Thus, water and wildlife are nonmarket resources not because of any failure in the marketplace but because governments have interfered with the market. The same is true for recreation, but the interference takes a different form. In the West, most outdoor recreation opportunities are on public lands. The Forest Service in particular provides more recreation than any other federal agency. Yet federal law prohibits the Forest Service from charging for most forms of recreation.

Private landowners can offer recreation such as camping, hiking, and picnicking. But most in the West and many in the East would have difficulty making money because they must compete with free or below-cost recreation on public lands. Private landowners are thus offered little incentive to maintain water quality or quantity, protect wildlife habitat, or provide recreation. Although this is primarily due to government interference with the market, it is ironically termed a market failure. Instead of trying to make the market work, reformers have sought increased government interference — which has often proven counterproductive.

By giving national forest managers the right — and the incentive — to charge recreation fees, marketization encourages the promotion of recreation uses. According to the Forest Service, recreation is more valuable than timber and other commodities in every region except Region 6 (figure 13.1). Managers today receive no sense of this value because only a small fraction of recreation fees are actually collected (figure 13.2) — none of which is retained by the managers. With the incentive of fee collection, forests will produce game fish and wildlife, provide primitive and semiprimitive recreation opportunities, and improve existing recreation facilities to increase their collections and their budgets. In turn, private landowners will also be encouraged to produce recreation and collect recreation fees.

Certainly, some resources cannot be bought and sold in a marketplace. Endangered species is the major example in the national forests. Although society may want to protect all species, contributions to an endangered species fund would necessarily be voluntary and would probably be insufficient to protect such species as the red-cockaded woodpecker and northern spotted owl. Prescriptive legislation may be needed in such cases. However, marketization can take the pressure from some endangered species, such as the grizzly bear, whose habitat is being damaged by below-cost activities.

Although marketization is not perfect, it will greatly improve the environmental quality and economic efficiency of national forest management. Few other systems will put an end to below-cost timber sales or below-cost grazing, and no other system will allow forest managers to respond rapidly to changes in demand. Marketization is the only system that guarantees the efficiency of forest management while providing safeguards for environmental protection.

Marketization Is Equitable

Economists distinguish efficiency from equitability — questions of the distribution of wealth and income. Compared to interest-group politics, where national forest resources go to those who have the funds to lobby Congress, the market is relatively neutral with respect to income distribution. Wealthy people build homes of cedar and redwood, while less well-to-do people rely on Douglas-fir and pine. Wealthy people are able to afford forms of recreation that require expensive equipment or guides, while lower-income people make do with lower-cost recreation activities.

Unlike the current system, marketization will not tend to allow the rich to get richer at taxpayers' expense. There will be no more appropriations for expensive roads to allow timber companies to profit from below-cost timber sales. Cattle grazing will not be favored over wildlife habitat for ranchers who are not willing to pay the full cost of range management. Of course, marketization cuts both ways: New wilderness areas will be created only where the wilderness values exceed timber values. But given the broad extent of below-cost timber sales, wilderness values will probably exceed timber values on millions of acres.

Although marketization will not make the distribution of wealth any worse than it is today, many people believe there are problems with the current distribution. In the past this argument has been used to justify many actions in the national forests — from below-cost timber sales and grazing to the Shelton Cooperative Sustained Yield Unit. Most of these actions ended in enriching a few at a cost to many.

For example, the Shelton Unit, on Washington's Olympic Peninsula, gave Simpson Timber Company exclusive rights to bid on timber in an entire Forest Service ranger district. The purpose was to protect the stability of the community of Shelton, Washington. The lack of competition cost taxpayers tens of millions of dollars per year in potential revenues in the late 1970s, yet no jobs were created by the unit that would not have existed somewhere in the vicinity.[4] Instead, the millions of dollars of revenue losses simply represented a transfer of wealth from U.S. taxpayers to Simpson Timber Company stockholders. Similar transfers on a smaller scale resulted from other sustained yield units in Washington, Oregon, New Mexico, and Arizona. David Clary describes the law that made sustained yield units possible as "the fruit of industrial support in Congress."[5] Such massive transfers of wealth would not be possible in a marketized Forest Service.

The Forest Service is not the appropriate agency for correcting inequitable distribution of income. The Department of Health and Human Services, which was created to specifically address this question, may be able to do so much more effectively than an agency created to manage forests. In the meantime, a marketized Forest Service will ensure that the distribution of national forest resources is far more equitable than it has been in the past.

Marketization Will End Polarization

In contrast to interest-group politics, marketization encourages cooperation rather than polarization. Under the current system, everyone has the opportunity to make substantial gains from transfers of wealth from the public to themselves. Such transfers are possible only if interest groups refuse to cooperate with other potential resource users since cooperation would reduce the amount of any particular wealth transfer.

By encouraging productive activities instead of wealth transfers, marketization will end such polarization. Timber companies will want to cooperate with recreationists because they can profit by encouraging recreation on their own land. Wildlife lovers will want to cooperate with ranchers because they may be able to increase wildlife habitat by paying ranchers to reduce domestic grazing. Marketization will make such cooperation desirable to all forest interest groups.

Objections to Recreation Fees

Of the major features of marketization, recreation fees may be the most controversial. Forest recreation researchers at the University of Idaho and Rocky Mountain Forest and Range Experiment Station list five common arguments against recreation fees.[6] These arguments, and appropriate responses to them, are:

1. "It is unfair to charge recreationists twice, once through taxation and once with user fees." But after marketization, no tax dollars will be spent on Forest Service recreation, ending any danger of having to pay twice. User fees are also more equitable than taxation since only those people who actually use the resource will have to pay for it.
2. "Fee programs are inequitable, because they discriminate against people who cannot afford to pay the fee." When the costs of recreation vehicles and other specialized equipment are summed, fees would represent only a small portion of the total and so would not be a major deterrent to recreation for the poor. Most recreationists are not poor, and it is unfair to use tax money to subsidize all recreation just to make it available to the relative few who might not be able to afford recreation fees. If fees are truly a deterrent to low-income people, some form of "recreation stamps" could be provided by the Department of Health and Human Services.
3. "Recreation resources should be subsidized with public funds because recreation is good for society as well as for the individual participant." There is no reason why this argument should be limited to outdoor recreation on public lands. Why should Disneyland, baseball games, movies, or concerts cost money if outdoor recreation is free? Taking this argument to its logical conclusion, the government should subsidize video games, dirt bikes, and bowling.
4. "Some form of subsidy is needed for future generations." This statement implies that the political system is better than the market at taking future generations into account. At a 10 percent discount rate, a recreation fee of $3 that can be collected ten years from now is worth more than $1 today to someone marketing that recreation, while at a 5 percent discount rate it is worth nearly $2. What is recreation ten years from now worth to a politician? U.S. Senators face reelection every six years, while representatives must campaign every two years. Presidents may

ten years is probably worth very little to any of these elected officials.

5. "Outdoor recreation [is a] relatively unstructured experience. The fee-collection process itself is then seen as degrading the leisure experience." This is belied by the large number of people who readily pay recreation fees today. Outfitters and guides typically collect $100 per day for hunting, fishing, boating, and horseback riding. Ski resorts collect fees from downhill skiers for lift tickets, and many western states collect fees from cross-country skiers for clearing snow from parking areas. A few private landowners in the eastern United States charge hikers an entry fee for backcountry hiking, while a growing number of ranchers in the West are charging hunters for permission to hunt. Hunters and anglers are already used to buying licenses, while fees are often collected for camping, swimming, picnicking, and other activities on both public and private lands.

Many of these same arguments were made when the Forest Service began charging fees for timber and grazing. Yet the arguments are no more applicable to recreation than they are to commodities. For example, the notion that recreation is good for society is equally, if not more, applicable to housing, food, or a number of other items. A society that allows a large segment of its population to remain unsheltered and unfed is likely to be much more unstable than one that charges fees for outdoor recreation.

At heart, the real objection to fees may be that national forests have traditionally (and by law) never collected fees for most recreation. Imagine that Disneyland had been free since it opened. Crowds and lines to attractions would be even worse than they are now. Disneyland would produce no profits, so many of the more recent attractions, not to mention Disney World, would never have been built. In addition, other businesses, such as Knott's Berry Farm and Magic Mountain, would have little incentive to compete with Disneyland. Finally, if the managers of Disneyland announced that they were going to start charging fees, people who believed they had a right to free use of Disneyland would strongly protest. People who are used to getting something for nothing will naturally be surprised and upset when it is suggested that they begin to pay. Yet fees are, in fact, in the best interests of recreationists.

Marketization Increases Recreation Opportunities

A major benefit of recreation fees is that they will prevent overuse of popular recreation areas. Higher fees in crowded areas will encourage some people to use other sites that, because they are less popular or farther from population centers, are less used and thus require lower fees (figure 13.3). Higher weekend fees for some forms of concentrated recreation will distribute recreation use to other parts of the week.

Recreation fees are more fair than the lottery or waiting list systems, which are used to allocate many forms of recreation today. Many states distribute hunting licenses on a lottery basis that may, for example, give all applicants a one-third chance of getting a permit in any given year. River rafting permits are often dis-

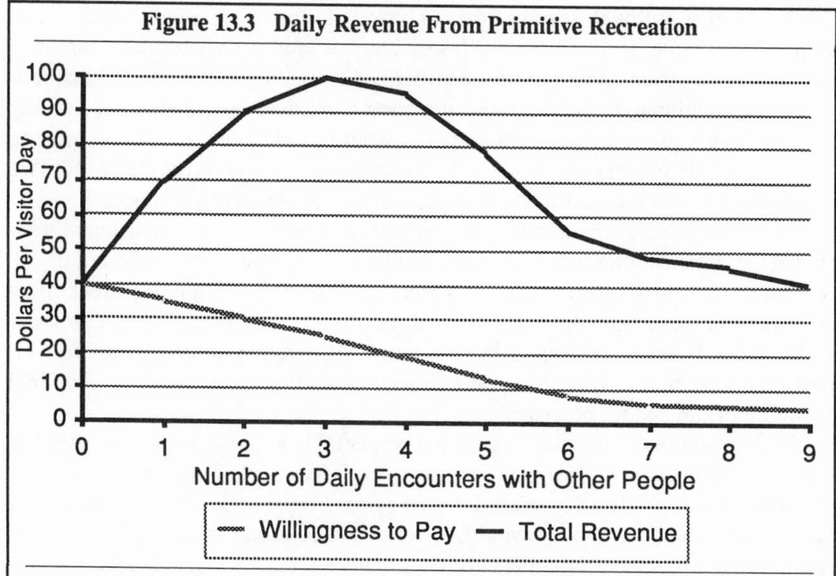

Figure 13.3 Daily Revenue From Primitive Recreation

People's willingness to pay for primitve recreation will vary with the number of encounters they have with other people. At higher fees, fewer people will use an area, so encounters will decline. Managers will tend to choose a fee level which maximizes the total revenue.

tributed by federal and state agencies on a similar basis. The chance of a private party getting a permit to run Idaho's Selway River, for example, is very slim. The National Park Service uses a waiting list to allocate private party trips on the Grand Canyon — and the list is about seven years long.

Such methods of distribution fail to ensure that recreation will go to those who value it most. Of two applicants in a state hunting lottery, for example, one may live for hunting and care little about other sports, while the other might find hunting only mildly enjoyable. Each has an equal opportunity to get a license, and because both apply along with many others, each gets a permit only once every three or four years. This is probably acceptable to the second applicant but very frustrating to the first. Under a fee system, each can hunt as often as their desire is strong enough to pay the fee, which in one case may be every year and in the other only every four or five years.

Recreation fees would also help reduce conflicts among recreation uses. Conflicts exist between motorized and nonmotorized forms of recreation, such as hikers and off-road vehicles, as well as between outfitters and private parties, particularly on rivers where the total number of parties that can start per day is limited. Allocations between uses would be based on willingness to pay. In practice, for example, hikers would be able to choose between using an area open to off-road vehicles or paying a premium to hike in areas with no off-road vehicles.

There would be little danger that certain forms of recreation would come to dominate entire national forests to the exclusion of other uses. Mechanized recreation, for example, tends to require fewer acres of land per visitor day than nonmotorized recreation, mainly because motorized recreationists tolerate higher densities of recreation use. For example, devoting more than a tiny percentage of any national forest to recreation vehicle parks would produce no additional income since the supply would quickly saturate the demand — but could actually sacrifice income from other recreationists who preferred to avoid developed uses.

The strongest argument in favor of recreation fees is that they will actually increase the opportunities available to recreationists over the present situation. As two Yale University forest economists recently observed, "While the direct effect of recreation fees is to reduce use, the indirect effects will increase recreation opportunities. Private investment in recreation will become more profitable. Increased investment in recreation on public lands will occur through the political economy of the budget process."[7]

Fees would make recreation a highly respectable activity to county officials, whose agencies receive 25 percent of all Forest Service receipts and who now focus almost exclusively on increased timber sales. Forest officials who overlooked potential recreation activities would be reminded of such opportunities by these county officials. Such reminders would usually be unnecessary since the forest managers themselves would depend on receipts for their own budgets.

Forests that now emphasize below-cost sales would quickly begin to promote recreation instead. Many forests whose timber is marginal would find that recreation is a better investment and that timber sales in at least some areas would reduce potential recreation income. Even forests that have valuable timber would mitigate the visual and other effects of timber sales in many areas to increase recreation income.

National forest recreation fees would affect recreation on other lands as well. With the elimination of a major below-cost competitor, private landowners would have the opportunity to charge for recreation. While not all may take advantage of such opportunities, many large landowners, such as timber companies and ranchers, would begin to collect fees for hunting and other popular recreation activities. In turn, they would have the incentive to modify their management practices to provide recreation experiences worth the fees they charge. In particular, wildlife habitat would be improved to attract hunters and water quality would be protected to maintain fishing potential. Some landowners might even promote hiking, camping, and other forms of recreation.

Collecting Recreation Fees

The debate over recreation fees would be moot if it were impossible to efficiently collect fees in the national forests. Collection booths at every national forest entrance, similar to those used by the Park Service, would be formidably expensive to maintain, simply because so many roads lead into and through the forests. Fortunately, other methods of collection and enforcement are available.

In a decentralized National Forest System, each forest would be able to experiment to find the best fee collection and enforcement systems. Most forests are likely to develop a multitiered fee structure that requires one or more of several possible permits for any recreation activity.

The basic permit would be a general dispersed recreation license which would be required for any forest entry off U.S., state, or county highways. The permit would be displayed as a sticker on car windows or bumpers and so would be visible to forest officials. An annual permit might cost $20 or so, and daily or weekly permits would also be available. Autos without permits could be stopped by forest officials, who would simply sell them a permit; in parking lots autos without permits could be tagged with something similar to a parking ticket.

Additional permits would be required for special activities. These would fall into at least three basic classes: hunting and fishing, wilderness, and concentrated use permits. Hunting and fishing fees would be the easiest to collect since state agencies already have an enforcement structure. When hunting or fishing licenses are purchased, licensees would state whether they plan to hunt or fish in a national forest. If so, they would pay a surcharge. The state would retain some of the money and turn the rest over to the national forest. Large private landowners could make similar arrangements with the states, whose budgets would also be enhanced by the fee structure.

A precedent for such hunting permits can be found in several eastern national forests. In Virginia, for example, hunters in the George Washington and Jefferson national forests pay a surcharge to the state wildlife department. Income from this surcharge is used for wildlife habitat improvements in the national forests.

Wilderness permits would be required for entry into any wilderness area. The cost of the permit would vary, depending on the popularity of the wilderness. Wilderness permits could be worn as tags on coats or backpacks, similar to the tags used for ski lift passes. Permits would be enforced by the system of wilderness guards, which the Forest Service already has in place in many wilderness areas.

Concentrated use permits would vary widely in price and method of collection. Concentrated uses include river trips, camping and picnicking, cross-country skiing, off-road vehicles, and many other forms of recreation.

A particular forest might choose to sublet campgrounds to private operators. The operators would be responsible for maintaining the campgrounds and collecting fees. A share of the fees would be paid to the forest on a royalty basis. Cross-country ski and snowmobile areas might be sublet to sporting clubs. Ski and snowmobile clubs could groom the trails, collect fees, and pay the royalty to the forest. River trips on the forest might be administered by the forest itself, which would specify fees and take reservations for a fixed number of trips per day.

Although the forests would be decentralized, they would be likely to cooperate in establishing fees and collection systems. All forests in one state or region, for example, might offer a basic dispersed recreation permit good for any forest in that region. Most other permits would be localized, making it possible for managers to receive direct feedback on recreation use and demand.

Several permits might be required for a given outing. A hunter, for example, may want to have a base camp in a developed campground and hunt from there into a wilderness area. The hunter would buy the basic dispersed recreation permit and wilderness permit from the forest and a hunting permit from the state (all of which would be sold by most sporting goods stores). Daily fees would be paid for campground use. While this may seem complicated, it is less complicated than the day-to-day life of Americans who pay for housing, utilities, food, and other goods and services to various entities.

While fees for a few premium recreation areas might be prohibitive to some people, most national forests enjoy an excess of total recreation capacity. This means that everyone will find affordable recreation opportunities in their area of interest. For example, premium river trips, such as on the Grand Canyon of the Colorado (were it under national forest jurisdiction), might sell for $100 per day — fees rivaling outfitter fees. Trips on other popular rivers, such as Idaho's Salmon, might sell for $25 to $50 per day — which is approximately equal to outfitter fees minus the costs of food and equipment. Trips on less popular rivers, such as Oregon's Illinois, might sell for as little as $10 per day.

Recreation Versus Commodity Uses

Relatively modest recreation fees would be needed to cover costs or to compete with commodity use of the national forests. The Forest Service estimates that in 1990 the total benefits of national forest recreation (including wilderness and wildlife recreation) will be over $2.7 billion per year (1987 dollars), or about $10 per visitor day.[8] Six percent of this amount would cover the total costs of the existing recreation and wildlife programs, which in recent years have cost about $150 million per year.[9] Another 4 percent should cover the costs of fee collection and enforcement. Since it is likely that more than 25 percent of the benefits could be collected as fees — which would mean an average fee of just $2.50 per day —the Forest Service's share would cover costs, and any additional collections could be used for expansion of the recreation and fish and wildlife programs.

The Forest Service also estimates that timber sales will return $1.6 billion in 1990 (1987 dollars). This estimate, however, is based on improbably high timber values in several Forest Service regions. For example, to meet Forest Service projections, timber sales in Alaska's Tongass National Forest would have to bring over $175 per thousand board feet. In reality, sale prices in that forest have rarely exceeded $50 per thousand board feet. Applying more realistic prices to proposed timber sale levels brings total returns to about $1.4 billion.

To produce revenues equal to timber, national forest recreation receipts need be little more than half of the Forest Service's estimate of benefits. Moreover, this is skewed by the very high timber values in Region 6, which produces nearly 40 percent of the volume and half of the value of timber in the National Forest System. Outside Region 6, recreation receipts need only be one-fourth of estimated benefits to exceed timber receipts. This ranges from $0.40 per visitor day in Region 2 to $5.70 in Region 8, the latter being the only region other than Region 6 where receipts would have to exceed $4 per day to equal timber receipts (figure 13.4).

Even in regions with valuable timber, such as the Northwest and South,

Figure 13.4 Recreation Fees Needed to Equal Timber Receipts

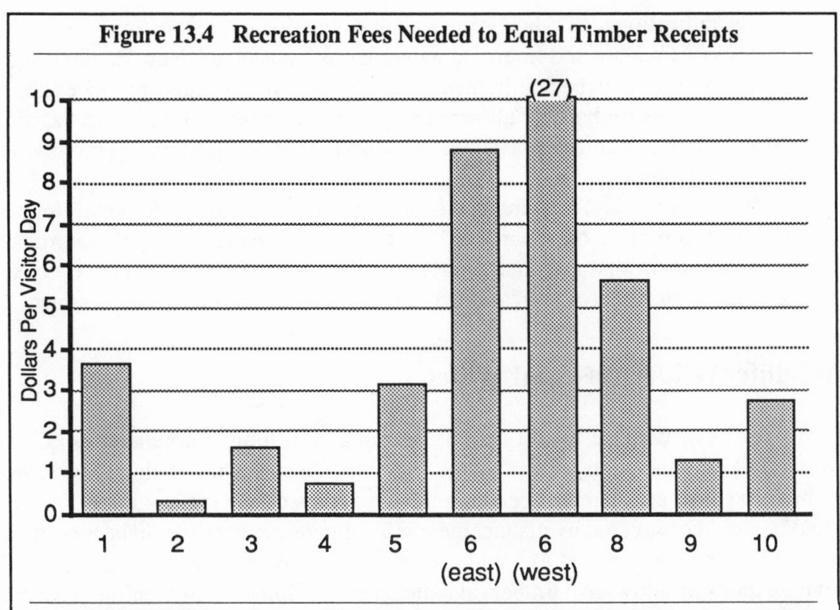

Average fees of $2 per day would make recreation highly competitive with timber in a majority of national forests. Average fees of $5 per day would bring recreation income above timber receipts in every national forest outside of the Pacific Northwest and the deep South.

Source: "Total Benefits – 1985 RPA High Bound Program," unpublished document on file at Forest Service office in Washington, DC, 1987, with modifications as explained in the text.

recreation would play an important role in management, in part because many popular recreation areas are in areas of low timber values. In the Southern Region, for example, the most popular recreation areas are in the Ozark and Appalachian Mountains, where timber values are so low that most sales are below cost. The highest timber values are in the Southern coastal plain forests, which are flat and relatively uninteresting to recreationists.

In Region 6, west-side forests contain 40 percent of the region's commercial forestlands yet produce 80 percent of the timber value.[10] In comparison, over 40 percent of the recreation use, and presumably recreation value, is found east of the Cascades.[11] East-side recreation receipts of $8.85 per day would exceed timber receipts, while west-side fees would have to equal $27 per day to exceed timber receipts. This suggests that timber will remain an important use on the west side, but recreation will effectively compete with timber on many of the less productive timberlands on the east side.The basic dispersed recreation fee would be fairly low —perhaps less than $1 per day. The real income would be generated from hunting and fishing permits and concentrated use fees. Most forests could produce fees averaging $3 per visitor day — which is less than a third of the Forest Service's estimated recreation values and in fact is the estimated value of the least valuable recreation in the national forests. At this rate, recreation would be the main source

At this rate, recreation would be the main source of national forest income throughout the Rocky Mountain and Intermountain regions, Alaska, the New England and Midwestern forests, and the Southern mountain forests. It would also produce about as much income as timber in California and northern Idaho. Only in the Pacific Northwest and the Southern coastal plain would timber be significantly more valuable than recreation.

At $3 per visitor day, total income from recreation will exceed $900 million in 1990. Counties receive one-fourth of this, greatly increasing their national forest income. Assuming that national forests are allocated two-thirds of net receipts, Forest Service budgets for recreation and wildlife would be more than doubled.

Wildlife, Wilderness, and Water

Many forms of wildlife would benefit from marketization. In general, wildlife would fall into three categories: Game and other species whose habitat needs are similar to game; endangered species; and all other species. Game animals, in particular, would be favored by management. Endangered species would be protected by the Endangered Species Act, and many would enjoy improved habitat because of reductions in below-cost timber sales and grazing. Some strengthening of the act may be needed to protect the spotted owl and other species threatened by the harvest of valuable old-growth timber. Populations of other species may increase or decrease, but as long as they did not approach threatened or endangered status, there would be little cause for concern.

Krutilla and Fisher have shown that markets may not provide an optimal amount of wilderness. This is partly because many people place an "option demand value" on wilderness — that is, they may not expect to use wilderness soon, but they wish to have the option to do so.[12] In addition, Krutilla and Fisher suggest that there are some "public goods" aspects of wilderness — many people would benefit from the existence of wilderness even if only a few people pay the costs.[13] Finally, Krutilla and Fisher show that, because a decision to develop a wilderness is irreversible, special caution must be exercised before any such decision is made.[14]

For these reasons, it may be necessary to provide some protection for wilderness and wild lands above that which would be provided by the market. At the same time, markets can play a role in allocating lands that are eligible for wilderness status but not yet so designated. For example, the wilderness system could be expanded by using the receipts from wilderness user fees to purchase development rights on other lands.

To protect and expand the wilderness system, all designated national forest wilderness areas, along with wild and scenic rivers, research natural areas, and other designated wild lands, should be placed in wilderness trusts that will be managed separately from, but in cooperation with, the national forests. The trusts would be administered by boards of trustees who would be obligated to meet the provisions of the 1964 Wilderness Act but who would be allowed to charge fair market value for permitted wilderness uses.

In many cases, the trusts would contract with the national forests to actually collect the fees. The forests, however, would receive only an administrative payment rather than two-thirds of net revenue. The remaining funds, along with private donations, would be used for maintenance and expansion of the wilderness system. The wilderness boards would have the right to purchase development rights on other national forestlands if they "outbid" developers by paying an amount in excess of the estimated value of the developed resources on those lands. They could also purchase development rights on private lands through negotiations with the private landowners.

Like wilderness, water and water-related resources may be only partially protected through markets. After marketization, the Forest Service would have an incentive to maintain the water quality of popular fish streams to produce income from the sale of fishing permits. Other streams would be protected by a decrease in below-cost timber sales and grazing. However, the Clean Water Act would need to remain in force to protect many other streams.

To increase the effectiveness of marketization, state water laws should be changed to allow the Forest Service to appropriate water and to allow all owners of water rights to transfer appropriated rights to other users. National forests could then sign contracts with downstream municipalities or other water users to provide quantities of water at a specified level of quality. One possible contract would have the user pay the forest if the terms of the contract are met, but have the forest pay the user if water quality falls below the specified level due to forest activities. Such contracts would provide powerful incentives to maintain water quality.

To summarize, marketization would improve the management of most forest resources, including timber, range, recreation, wilderness, most species of fish and wildlife, and, to a lesser degree, water quality. To protect resources that the market might not adequately produce, the Endangered Species Act, the Clean Water Act, and the Clean Air Act should remain in force.

Legislation Needed to Overcome Bureaucratic Inertia

Eighty years of bureaucratic inertia render the Forest Service incapable of making the changes neeeded to correct the environmental and economic problems of the national forests. Reform must come from outside the agency, and it must dramatically change the institutions and bureaucracy under which the national forests are managed.

If, as this book suggests, the Forest Service is motivated primarily by its budget, then reforms must be aimed directly at the budget. New incentives can be created to encourage recreation where recreation is most valuable and to encourage efficient timber management on the most productive timberlands by making the agency's budget sensitive to the demand for these resources. The next chapter details the specific changes needed in the law to achieve these reforms.

Notes

1. Paul Kennedy, "What Gorbachev Is Up Against," *Atlantic Monthly* 259(6):29–43.
2. Peter Drucker, *Innovation and Entrepreneurship* (New York, NY: Harper & Row, 1985), p. 185.
3. USDA Forest Service and USDI Bureau of Land Management, *Grazing Fee Review and Evaluation* (Washington, DC: Forest Service and BLM, 1986), p. 40.
4. Surveys and Investigations Staff, Committee on Appropriations, U.S. House of Representatives, "Timber Sales Process of U.S. Forest Service" (Washington, DC: House of Representatives, 1982), p. 41.
5. David A. Clary, *Timber and the Forest Service* (Lawrence, KS: University Press of Kansas, 1986), p. 126.
6. Charles C. Harris and B. L. Driver, "Recreation User Fees: Pros and Cons" *Journal of Forestry* 85(5):25–29.
7. Clark S. Binkley and Robert O. Mendelsohn, "Recreation User Fees: An Economic Analysis" *Journal of Forestry* 85(5):31–35.
8. "Total Benefits – 1985 RPA High Bound Program," unpublished document on file at Forest Service office in Washington, DC, 1987.
9. USDA Forest Service, *1987 Budget Explanatory Notes for Committee on Appropriations* (Washington, DC: Forest Service, 1987), p. 101.
10. Florence K. Ruderman and Richard W. Haynes, *Volume and Average Stumpage Price of Selected Species on the National Forests of the Pacific Northwest Region, 1973 to 1984* (Portland, OR: Forest Service, 1986), p. 8; Randal O'Toole, "Are Region 6 Forests Being 'Overcut'?" *Forest Watch* 7(10):5; USDA Forest Service, *Draft EIS for the Pacific Northwest Region Plan* (Portland, OR: Forest Service, 1981), p. 17.
11. USDA Forest Service, *Regional Guide for the Pacific Northwest Region* (Portland, OR: Forest Service, 1981), pp. 3-18 to 3-36.
12. John V. Krutilla and Anthony C. Fisher, *The Economics of Natural Environments* (Baltimore, MD: Johns Hopkins University Press, 1975), pp. 14–15.
13. Ibid., pp. 23–24.
14. Ibid., pp. 41–43.

Chapter Fourteen
Specific Proposals

Ten specific proposals for marketizing the Forest Service are presented in this chapter. For the reforms to be completely successful, all should be implemented. All but one of the reforms can be implemented by an act of Congress.

Proposal One: Eliminate Improper Incentives

The laws that encourage the Forest Service to act inefficiently should be repealed. These include the Knutson-Vandenberg Act of 1930, the Brush Disposal Act of 1916, and the portions of the National Forest Management Act relating to timber salvage sales. In addition, the Public Rangeland Improvement Act of 1978, which dedicates half of all grazing receipts to range improvements, should be repealed because it offers improper incentives for grazing similar to those the K-V Act provides for timber.

Proposal Two: Provide New Incentives

Operations and maintenance of each resource on each national forest should be funded out of a portion — probably about two-thirds — of the net receipts produced by that resource. This will give managers the incentive to maximize net returns from national forest management. The formula giving counties 25 percent of gross receipts would be retained, and the U.S. Treasury would collect the rest. Since two-thirds of net receipts is equal to 40 percent of gross receipts, this means that the Treasury would collect about 35 percent of receipts. Of course, allocating two-thirds of the net, rather than 40 percent of the gross, to management gives managers an incentive to maximize net rather than gross receipts.

Proposal Three: Revise Fee Collections

Fees at fair market value should be collected for all resources. Rather than legally defining fair market value, local managers should be allowed to determine fees based on available information. Given that their funding will come out of receipts, managers can be relied upon not to sell resources below cost.

Proposal Four: Eliminate Appropriations

Given that fees can cover most operations and maintenance, there is no need for Congress to continue to appropriate funds for most Forest Service activities. Only a few budget items would require continued funding. Yet if any funds are appropriated by Congress, a budget-maximizing agency will seek to expand those appropriations by expanding its authority into new areas. As a rule, then, no funds should be provided by Congress to the Forest Service. Those resources or activities that require additional funding should be funded through other agencies.

Only three such activities appear to exist, and these only in limited circumstances: threatened and endangered species, watershed protection, and research. Funds for these activities should be appropriated to the Fish and Wildlife Service, Soils Conservation Service, and National Science Foundation.

Many problems with threatened and endangered species will be greatly reduced by reductions in below-cost timber sales. National forests will still be required by the Endangered Species Act to protect habitat for rare species even where timber is valuable. In a few instances, habitat improvement activities, which would not normally be paid for by national forest wildlife receipts, may be needed. These activities could be done by contracts between the Fish and Wildlife Service and the national forests.

Similarly, reductions in timber and grazing activities should eliminate many watershed problems. National forests will still be required to obey the Clean Water Act. The desire to produce income from fishing and, possibly, municipal or industrial water users will encourage forest managers to maintain water quality. In a few instances, however, high timber values or past practices may create watershed problems that require rehabilitation. Such activities could be accomplished by contracts between the Soils Conservation Service and the national forests.

Most research now done by the Forest Service benefits some market or marketable resource, including research on timber, recreation, grazing, and many forms of wildlife. Such research can continue through contracts between the experiment stations and the national forests or other entities marketing those resources. A minority of research projects, such as those on long-term forest productivity, nongame wildlife, or other areas, may be important as basic research yet would not normally be funded by market-oriented national forests. Such research should be done by contracts between the National Science Foundation and the national forests or experiment stations.

While fees may cover all other forms of operations and maintenance, capital expenditures require additional funding. Rather than use appropriations for such funding, the Forest Service should be authorized to sell bonds — perhaps tax-free bonds, like those of other public entities — to raise money for capital improvements. Funds from the sale of bonds could be used for roads, reforestation, recreation facilities, and range and wildlife habitat improvements. Repayment of the bonds would be made out of fee collections. Naturally, this system would encourage the Forest Service to use a more reasonable discount rate when considering future investments.

Proposal Five: Decentralize the Forests

Because all Forest Service operations will be paid out of user fees, there will be little need and less financing for the huge, thousand-person staff in the Forest Service's Washington office. Many of the three thousand people in the nine regional offices will also be unnecessary. In place of this centralized administration, each national forest should be chartered as an independent, publicly owned (in the sense that the Post Office is publicly owned) organization.

The forest supervisor of each national forest would essentially be the chief executive officer. The supervisor would be responsible to a board of directors consisting, possibly, of two people appointed by the governor of the state in which the national forest is primarily located and three people appointed by the secretary of agriculture. These boards would determine policies for each national forest. Given the fact that incentives have a greater long-term influence on organizations than people, the actual method of appointing board members is less important than the method of financing national forests.

The charters would give national forests broad authority to sell or lease timber, recreation, grazing, and other forest goods and services. The forests could do this themselves or enter into contractual relationships with other organizations. To maintain flexibility, contracts should be limited to no more than ten years in length and should be fully transferable.

A national forest might choose any of a number of ways of managing resources. For example, one forest may choose to monitor and collect recreation fees itself. A second forest might contract with nonprofit recreation groups to provide certain forms of recreation such as cross-country skiing, hiking, or boating. Another forest might contract with for-profit organizations to provide similar services. This diversity will ensure that many management techniques are tested. Ultimately, the most successful will be adopted by more forests.

If regional and Washington offices of the Forest Service continue to exist at all, it would be mainly to provide support services for the national forests. Research, forest survey, and certain other tasks might be more efficiently undertaken by regional or national offices and paid for by subscriptions from the local offices. However, these offices would have to compete with private firms, which might be willing to provide the same services at lower cost.

The State and Private Forestry branch of the Forest Service could continue to exist to provide low-interest loans to private forest landowners. This practice would stimulate improved management of private timberlands, which, on the average, are much more productive than the national forests. Funding for the loans would come from the sale of bonds, which presumably pay a lower rate of interest than other loans that could be obtained by private landowners.

In the event that a chartered national forest is unable to pay expenses out of its share of income — goes bankrupt — adjacent national forests and the local wilderness trust would be given the opportunity to bid on its lands. Successful bidders would be required to pay all debts and to respect existing contracts. Lands that are not bid upon would be managed on a custodial basis.

Proposal Six:
Create Wilderness Trusts

Wilderness areas represent a unique resource that should be managed using a special system. All wilderness areas in a given region should be overseen by a board of trustees that is obligated to obey the terms of the Wilderness Act and to maximize wilderness values. Fifteen separate trusts and boards should be established, one for each of the twelve western states, Minnesota, the remainder of the Midwest and Northeast, and the Southeast.

Wilderness recreation fees collected by the trusts (or by the national forests on behalf of the trusts) would pay for wilderness operations and maintenance. Wilderness expansion would also be possible by allowing wilderness boards to purchase development rights on other lands in the national forests. Recreation fees might not be sufficient to permit rapid expansion, but additional income could come from donations, just as today donations fund the preservation of private lands through such organizations as the Nature Conservancy. Funds would also be available from the sale of other resources in the wilderness, including grazing permits and minerals, at the discretion of the wilderness boards.

For example, an oil company might propose to extract oil from a wilderness area using methods that require no surface occupancy. Royalties from this extraction could be used by the wilderness trustees to purchase timber rights from national forestlands adjacent to the wilderness. Total wilderness values might be increased if the acres added through purchase of development rights exceeded the acres disturbed by oil extraction.

Proposal Seven:
Compensate Displaced Workers

A large reduction in the level of timber sales would have a significant effect on employment in many localities, one that would not be immediately offset by increases in recreation jobs. A few jobs might also be lost in the grazing industry if ranchers are not willing to pay fair market values for current levels of grazing and can find no alternative source of feed.

Estimates in chapter 15 indicate that this proposal might directly displace as many as thirty thousand workers in the timber and livestock industries. In addition, Forest Service budgets — and therefore staff numbers — might fall by close to 33 percent, or about ten thousand people. Many of these jobs will be replaced by jobs in the recreation industry. However, many of the displaced workers will need to be trained in new jobs or placed in new locations.

Chapter 15 will also show that a "seed money" fund of $100,000 per displaced worker could be created with about 15 percent of the annual appropriations that this proposal will save. This fund could be provided to the national forests according to a formula predicting estimated job displacements. Although the fund would be specifically dedicated to compensating displaced workers, the forests would be given broad discretion over the use of the seed money. Some might subsidize below-cost sales during a transition period, others might provide low-interest loans to start new industry, and some might simply make payments to people for education and relocation.

Proposal Eight: Give National Forests Authority Over Subsurface Resources

Under current laws, the Bureau of Land Management has jurisdiction over most national forest subsurface resources. That jurisdiction should be given to the individual national forests. Jurisdiction over subsurface resources in wild lands should be given to the wilderness trusts. The forests and trusts should be allowed to charge fair market value for all such resources.

Proposal Nine: Repeal Unneeded and Meaningless Laws

Given the changes in national forest management just described, forest plans as conceived by the Resources Planning Act and National Forest Management Act are cumbersome and irrelevant. While independently chartered national forests will no doubt prepare management plans, there is no need for the plans to be as elaborate or as centralized as envisioned by these two laws.

Instead, the plans will be oriented to identifying and making the best use of opportunities to generate income through sales or leases of timber, recreation, and other resources. Public involvement in the plans will be in the form of proposals to use, lease, or contract parts of each forest for various purposes. For example, a hiking organization might propose to maintain trails, collect fees, and monitor recreation in a particular area in exchange for paying the forest a share of the receipts.

Along with RPA and NFMA, most other laws relating to national forest management should be repealed or amended so that they do not apply to the national forests. These would include the National Environmental Policy Act (NEPA); the Multiple-Use Sustained-Yield Act, a meaningless law that the U.S. Court of Appeals for the Ninth Circuit once said "breathes discretion at every pore";[1] the Forest Roads and Trails Act, which creates the purchaser credit system; the Land and Water Conservation Act, which makes collection of recreation fees illegal other than for developed recreation; and the 1872 Mining Act, which allows free use of federal minerals. Of course, these changes would not affect existing contracts and obligations.

Three laws that would be retained would be the Endangered Species Act, the Clean Water Act, and the Clean Air Act. The Endangered Species Act may need to be strengthened to ensure the protection of rare species of plants and wildlife, which may have little market value. As a corollary, provisions of NEPA should remain in force for national forest actions that may affect threatened or endangered species or lead to listing of a species. The Clean Water and Clean Air acts should apply to the Forest Service just as they apply to other landowners. Other laws that also apply to private land managers, such as laws regulating pesticide use, will also remain in force.

Many people will be shocked at the idea of exempting the Forest Service from NEPA. As far as the national forests are concerned, however, the most this law has done for the environment is to delay environmentally destructive projects. The projects were never permanently halted by NEPA. Most destructive projects contemplated by the Forest Service will be halted as soon as the agency is required to fund itself out of its fees. The few projects that remain might be stopped, if they can be stopped at all using court action, using the Endangered Species Act or Clean Water Act.

Proposal Ten: Change State Water and Wildlife Laws

Water resources can be protected best if the national forests — as well as other landowners — have an incentive to protect the waters that flow across their jurisdiction. No such incentive exists under current water laws. State water laws should be changed to allow landowners with water rights to sell those rights to others. The transfer of such rights will make it possible for fisheries groups to ensure minimum stream flows, for water that is now being used for irrigation to be used by municipalities, where such use is more valuable, and for landowners to have an interest in keeping waters clean.

Laws relating to fish and wildlife in some states also interfere with property rights and discourage environmental protection. In Montana, landowners are required by law to allow access to people who want to fish. Although designed to encourage fishing, this law perversely gives landowners an incentive to pollute streams so that fish populations are reduced and the landowners won't be bothered by anglers. In Wyoming, landowners are not allowed to collect fees for hunting. Again, the law gives landowners an incentive to destroy, rather than protect, wildlife habitat so they won't have to deal with hunters.

Comparison with Similar Proposals

These proposals are not entirely new. For example, a proposal similar to proposal five has been made by Dennis Teeguarden, a forest economist at the University of California at Berkeley. Under his proposal, "each national forest would become an independent public corporation, operating under federal charter with the legal authority to manage federal land in much the same manner that states charter private investor-owned public utilities."[2] Teeguarden's proposal differs from this one in at least two important details, which are worth discussing.

Teeguarden proposes that each national forest be required to pay the federal government "rent" equal to 11 percent of the asset value of the forests. Up to half of any remaining income would be used to manage the forests and the remainder would go to local governments. This differs from proposal five, which would leave the county revenue formulas undisturbed, provide two-thirds of net income to forest managers, and turn the remainder over to the federal Treasury. Under this arrangement, the Treasury would typically collect about 35 percent of gross income.

Teeguarden's proposal might provide more efficient management if "asset value" could be properly defined. Such a definition may be impossible, however, since the true value of recreation, wildlife, and other nonmarket resources is not yet known. Teeguarden's proposal would probably result in lower operating funds for the national forests as well as lower revenues to counties. For example, if Clawson's asset value estimate of $42 billion for the entire National Forest System is correct, 11 percent rent payments would require forests to pay $4.6 billion to the national Treasury each year.

Such payments could be achieved only by rapidly liquidating national forest timber inventories. To prevent such liquidations, Teeguarden proposes that each national forest would be required to "maintain sufficient growing stock to utilize at least 75 percent of the potential site productivity." Given the high asset value of some national forest timber inventories, many forests would be unable to meet this restriction and still produce an 11 percent rate of return.

Teeguarden also proposes creation of a "public corporations board" to regulate the national forests and "assure some reasonable degree of uniformity in corporate structure, operating policies, financial management, pricing policies, and resource stewardship." Given that diversity through decentralization of decision making is one of the criteria for reforms identified in chapter 13, such a board appears to be more dangerous than useful.

Teeguarden comments that "obviously, a profit-oriented public corporation would have no incentive to provide the level and set of non-market services that the public has come to expect [such as] access to recreation and wilderness or preservation of unique landscapes." He suggests that such services could be provided, "if judged in the public interest" by having government "agencies or private conservation groups enter into contractual agreements under which the public corporation is paid to provide specific services or to designate a particular area for a particular use."

Such contractual agreements are certainly one option for providing nonmarket services — yet they are not the only option. As described in chapter 13, national forest recreation values are high enough that profit-oriented managers would find it worthwhile to maintain scenic vistas, wildlife habitat, and a wide range of recreation opportunities in order to collect recreation fees.

Marion Clawson, of Resources for the Future, has suggested that national forestlands be leased with a "pull back" option. Under this system, anyone may apply to lease land for any purpose, but other competing users could lease up to a third of the original lease for their use under the same terms. For example, if a timber company applied to lease 90,000 acres of forestland, a wilderness group could apply to pullback 30,000 acres. Provided that the wilderness group paid the same same price per acre as the industry group, its pullback application could not be refused.[3]

Pullbacks are an innovative solution to current public land problems. However, they are not necessary to these proposed reforms. As long as chartered national forests can sell or lease resources and those leases are fully transferable, most resource uses should be protected and be available at efficient levels. Exceptions to this statement are limited to wilderness, threatened and endangered species, and possibly clean water. Wilderness would be protected by proposal six, and endangered species and water quality would be protected by existing laws.

Moreover, pullbacks are specifically contrary to the spirit of cooperation that these reforms are designed to generate. Under the pullback system, national forest users will have an incentive to lease more land than they really need to avoid losses through pullbacks or even to lease land before they expect to need it so that other users do not have access to that land. In a marketized system, users may sign transferable contracts with national forests. If another use is more valuable than the contracted use, groups supporting that use will be able to buy part or all of the contract. The contract owners should be willing to sell because the higher price will increase their profits.

The idea of wilderness trusts was first suggested by Richard Stroup and John Baden.[4] The fifteen wilderness trusts described above are a slight modification of their original proposal.

Partial Implementation

Many people may wish to consider partial implementation of these proposals for political, experimental, or other reasons. Such partial proposals might include marketizing selected national forests on an experimental basis; changing incentives (proposals one, two, and three) without changing the appropriations process (proposal four); or maintaining a centralized authority (proposal five) or a centralized planning process (proposal nine). A close look at each of these shows that the disadvantages of partial implementation outweigh the advantages.

Teeguarden suggested that his proposal to independently charter national forests could be tried in one region or a few national forests on an experimental basis. Applying marketization on such an experimental basis would put recreation at a particular disadvantage. If, for example, all forests in Region 1 (Montana and northern Idaho) were marketized, recreationists — particularly those in Idaho and southern Montana — would tend to use forests in Wyoming, eastern Washington, and southern Idaho to avoid paying recreation fees. This would give forest managers incorrect signals about the value of recreation on their forests.

If an experiment is to be conducted, Region 1 — which ranges from submarginal timber forests in eastern Montana through marginal timber forests in western Montana to some valuable timber forests in northern Idaho — might be a good place to conduct it. The results from such an experiment would have to be evaluated carefully to account for free recreation in adjacent regions. An even better region might be Region 8 (the Southern Region), which also includes both submarginal and supermarginal timberlands but does not have many nearby forests that could still provide free recreation.

People considering this marketization proposal will be tempted to maintain appropriations for at least some Forest Service activities. Certainly, proposal one — elimination of the K-V Act and other misincentives — is a worthwhile goal by itself, while proposals two and three — creation of new incentives by charging market fees for almost all resources and returning a share of the net to managers — will greatly improve many aspects of national forest management.

Eliminating appropriations, however, is important for the complete success of marketization. As long as appropriations are maintained, commercial interests will have an incentive to exploit the national forests by lobbying for resources below cost, while members of Congress will have an incentive to give in to such pressure to gain support in future elections, and the Forest Service will encourage such activities to enhance its own budget. The inevitable results will be inefficient management, large wealth transfers, and reduction of environmental quality. While proposals one, two, and three by themselves will result in significant improvements in national forest management, proposal four will insure that those improvements are not subverted by a short-sighted political process.

Suppose marketization were implemented with the provision that appropriations would be allowed only to protect wildlife. This would give the Forest Service and commercial interests the incentive to selectively quote research indicating that wildlife would benefit from below-cost timber sales. Appropriations would be made for road construction, pesticide application, and intensive timber management all in the name of wildlife.

Similar objections exist to the possibility of maintaining a centralized bureaucracy or at least a centralized, environmental impact statement-oriented planning process. A centralized bureaucracy or planning process could probably not be funded out of user fees and would require appropriations. The budget-sensitive bureaucracy would use these appropriations as a tool to gain further appropriations.

For example, a traditional job of the Forest Service has been to regularly inventory timber volumes on all public and private lands in the United States. These inventories were used in the RPA planning process and earlier documents to project future timber supplies. Invariably, the Forest Service uses the inventory data to project an imminent timber famine that can be prevented only by increasing national forest timber budgets.

Marketizing a single national forest region, such as Region 8, may still be a viable experiment. However, given the potential benefits to the environment and the economy, marketization should be implemented in the entire National Forest System as soon as possible. The results of such an action can be evaluated and, if successful, can be used to guide the marketization of other federal agencies.

Marketizing Other Agencies

Most other natural resource agencies could also benefit from marketization. At the very least, these would include the Bureau of Land Management (BLM), the Bureau of Reclamation, and the Corps of Engineers. Grazing rights on BLM lands are provided for $1.35 per animal unit month, just as they are in the national forests. The annual losses from this program are $50 to $100 million per year. Yet less than 10 percent of the cattle and sheep produced in this country are grazed on either Forest Service or BLM lands. Marketization of BLM grazing would improve the overall efficiency of red meat production.

The Bureau of Reclamation and Corps of Engineers each build dams and other water projects for irrigation, hydroelectricity, and other purposes. As John Baden notes, "Private irrigators pay only a small portion of the costs of the water" they buy from the bureau or corps. Since water is underpriced, their political associations "generate political demands for more dams than are socially optimal. Thus, in a period of huge crop surpluses, the Bureau of Reclamation is still proposing the development of dams to produce more new irrigated farm land. In addition to harming all taxpayers, existing producers of crops are injured."[5] These agencies can be marketized by allowing them to charge fair market value for the resources they produce and requiring that all their costs be paid out of user fees.

Complete proposals for marketizing other federal agencies are beyond the scope of this book. However, it is clear that many other agencies, particularly natural resource agencies, could be marketized. Such marketization would not only reduce federal deficits but also greatly reduce the environmental destructiveness of those agencies.

Marketization Can Begin Now

A single act of Congress implementing nine of the ten proposals in this chapter would greatly improve the management of recreation, wildlife, and other amenity resources as well as stabilize the supply of timber, forage, and other commodity resources. Proposal ten will require action by state legislatures, and water and some wildlife resources may receive less than optimal management until the legislatures act. Even before the states take action, however, management of these resources should still be improved over current conditions.

The specific proposals in this chapter are all subject to modification and improvement. To truly reform the Forest Service, however, the basic principles of marketization should be retained: Fund all activities out of net user fees; reduce appropriations to zero; charge fair market value for all resources; and decentralize the National Forest System.

Notes

1. *Perkins v. Bergland,* 608 F. 2d 803, 806 (9th Cir. 1979).
2. Dennis E. Teeguarden, "A Public Corporation Model for Public Land Management," presented at the Conference on Politics Versus Policy: The Public Lands Dilemma, Utah State University, Logan, UT, 23 April 1982.
3. Marion Clawson, *The Federal Lands Revisited* (Baltimore, MD: Johns Hopkins University Press, 1983), p. 216.
4. Richard L. Stroup and John Baden, *Cato Journal* 2(3):691-708.
5. John A. Baden, "Do We Need a Land Ethic?" presentation at the Liberty Fund Conference on Land Ethics, Big Sky, MT, June 1987.

Chapter Fifteen
Effects of Marketizing

The proposals in this book would clearly have dramatic effects on the Forest Service and the National Forest System. This chapter will attempt to predict the effects on U.S. and county treasuries, the outputs of timber and other resources, and forest-related employment.

National Forest Receipts

Congress currently appropriates $1.6 billion per year to the Forest Service but collects little more than $1 billion in receipts. Under this proposal, appropriations will fall to negligible amounts. Yet returns to the Treasury might actually increase.

Estimates of future timber receipts can be made from RPA estimates of benefits, which are broken down by region and resource.[1] Benefits are projected for 1986, 1990, and 2030 for water, minerals, range, recreation, wilderness, timber, and wildlife and fish in 1982 dollars. Unfortunately, some of the benefits appear unrealistic; for example, Alaska timber is projected to be worth over $175 per thousand board feet in 1990, while grazing values in the East are projected to be worth $71 per AUM. For this and other reasons, a number of adjustments to the benefits are necessary to estimate receipts under marketization. These estimates, in 1987 dollars, assume that proposed reforms are immediately passed by Congress and project the effects on receipts in 1990.

Although timber outputs will decline, elimination of cross-subsidies should result in an increase in timber receipts from forests that have many cross-subsidies. Judging from figure 8.5, this increase is likely to be at least $50 million per year. Forests that currently lose money on most of the timber they sell, such as those in Colorado and Alaska, will have an absolute reduction in timber receipts, but these receipts are small compared to the national total.

To estimate changes in timber receipts, the RPA values are adjusted to more realistic timber prices. For example, Alaska timber is assumed to be worth $35 per thousand board feet rather than $177. Smaller adjustments are made to values of other regions, and no adjustments at all are made to values in Regions 5, 6, and 8. Revenue increases due to an end to cross-subsidies are assumed to balance decreases due to an end to below-cost sales in Regions 1, 5, 6, and 8. An end to below-cost sales is assumed to reduce revenues in the other regions by $80 million per year. Values are also converted to 1987 dollars using a gross national product (GNP) price inflator, an increase of approximately 15 percent over the values in 1982 dollars. The result is that 1990 timber receipts are assumed to be close to $1.4 billion per year.

The Forest Service should be able to collect a significant fraction of the recreation benefits estimated by Forest Service economists. Forest managers will be able to practice "price discrimination" — charging higher fees to people who are willing to pay for premium recreation experiences. To be conservative, the RPA values are reduced by two-thirds to account for consumer surplus (the amount some people are willing to pay over and above actual fees), inability to collect some fees, and reductions in use in response to fee increases. Values are also converted to 1987 dollars using a GNP price inflator.

If forests can capture only a third of the recreation values claimed by the Forest Service, recreation receipts will greatly exceed combined timber and grazing collections everywhere except in Regions 5, 6, and 8. Even in those regions, recreation values will exceed commodity values in such places as the mountain forests of the South and southern California. Nationally, recreation and wildlife receipts will total over $900 million per year, as opposed to the $30 million now being collected.

Grazing receipts will probably also increase. Assuming that fees are raised to an average of $5 per AUM but that a fifth of ranchers will decline to pay those fees, grazing receipts should increase by more than three times — a movement from $13 million to nearly $40 million per year. No price inflator is used because grazing values are dependent on the price of meat, and meat — like most agricultural products — has not been increasing in value as fast as the GNP index.

Immediate receipts from water will be tiny because state legislation is also needed to collect receipts for most water. Even after such legislation is passed, water receipts will probably not exceed a fraction — perhaps 10 percent — of the estimated RPA benefits. This is because of the difficulty of collecting receipts for all water produced by a forest and because Forest Service estimates of water values are probably too high. At 10 percent of RPA benefits, water receipts will equal about $10 million per year. Although this is a small fraction of total receipts, water values could be significant in local municipal watersheds where the Forest Service and municipalities make management agreements.

The Forest Service estimates that 1990 mineral benefits will approach $900 million per year in 1987 dollars. However, few of these benefits are presently collected as receipts: 1986 receipts were about $158 million, and 1990 receipts are projected to be $250 million. The proposed reforms will not affect existing mineral rights under the 1872 Mining Act and other laws, but new mining activities will be required to pay market value. With these opportunities, actual receipts might reach $300 million per year by 1990.

Total receipts, then, will increase to about $2.75 billion per year by 1990. Most counties, as well as the U.S. Treasury, will gain from this proposal. Since counties collect 25 percent of total receipts, their share of annual receipts will increase from under $500 million in 1986 to nearly $700 million in 1990. Counties in all Forest Service regions will gain. After the Forest Service retains its two-thirds share of net income, receipts to the Treasury will exceed $950 million, more than three times the actual revenues, net of county payments, paid to the Treasury in 1986. Since two-thirds of net receipts is approximately equal to 40 percent of gross, the Forest Service's budget will fall from its current level of about $2 billion per year to about $1.1 billion.

Resource Outputs

The total sustained yield of timber will decline dramatically as below-cost and cross-subsidized timber sales are eliminated. As noted in chapter 2, the Forest Service estimates that 22 percent of its timber was sold below cost in the late 1970s. Chapter 9 indicated that up to 40 percent of national forest timber is appraised at negative rates, and a large portion of this timber is cross-subsidized. Assuming that there is some overlap between these two figures, an end to below-cost and cross-subsidized sales might reduce the economically sustainable level of timber sales by 30 to 40 percent.

This reduction will probably be phased in over several years as forests adjust to the proposed changes in Forest Service organization. Many forests, particularly the old-growth surplus forests on the west slope of the Cascade Range, might depart from the nondeclining-flow policy. Although such departures will be mitigated by the desire to maintain wildlife and recreation values and thus boost revenues from those sources, the departures should help to cushion decreases in national forest timber sales.

Forests in the Rockies and other submarginal areas will experience a major decline in sales. However, some forests in the Pacific Northwest, including northern California and northern Idaho, may increase sales as the old-growth surplus from lands dedicated to timber is cut at an economically optimum rate. This proposal will not necessarily eliminate timber sales below cost which the Forest Service claims are vitally needed for other resources. If the claims are valid — for example, if a sale is needed to protect or improve wildlife habitat — then wildlife managers could offer to subsidize the sale out of hunting receipts.

Grazing outputs will probably decline as range managers and ranchers adjust to fair market values for receipts and make investments only when costs are covered by receipts. Recreation use may also change in response to fee increases. As high fees will be charged for heavily used areas, people will be encouraged to use other sites, thus reducing the environmental effects of heavy recreation use.

Fees collected for recreation and wildlife-oriented recreation will greatly improve the quality of recreation in many areas. Recreation facilities will be improved, while fish and wildlife habitat improvements will increase the numbers of fish and wildlife available for consumptive and nonconsumptive use.

Employment

A 40 percent reduction in the level of timber sales will have a significant effect on employment in many localities, one that will not immediately be offset by increases in recreation jobs. Timber sales on most forests require between four and eight person-years of processing per million board feet. Assuming an average of six, and assuming that timber sales fall by 5 billion board feet, this proposal might directly displace 30,000 workers. With the Forest Service's budget falling by one-third to one-half, an additional 10,000 to 15,000 Forest Service employees will lose their jobs.

Since the proposal will save U.S. taxpayers $1.6 billion of appropriations per year and will return at least $600 million more than is currently being paid to the Treasury, it seems reasonable to use a part of this savings to assist displaced workers. At 7 percent interest, the capitalized value of $2.2 billion is $31 billion. A small fraction of this — about 13 percent — will provide a "seed fund" equal to $100,000 for each of 40,000 displaced workers.

This fund should be made available to local forests specifically for the benefit of displaced employees. The actual use of the funds should be decided by the national forest boards of directors. In some cases, forests might want to temporarily subsidize timber sales during a transition period for local communities. In other cases, funds could be used for job retraining and possible relocation of families. In still other cases, the funds could provide low-interest loans or grants to start new businesses.

Commodity Users

Mills that purchase a large share of their timber from below-cost timber forests will lose unless they were willing to significantly increase the amount they bid for stumpage. Mill workers in these areas will also lose, but these losses will be compensated by the proposed seed money fund. Mills that purchase from forests that have a large amount of cross-subsidized timber sales may suffer from less available volume of timber but not necessarily lower profits. Because cross-subsidized sales produce no more profits for the purchaser than sales of lower volume that include only valuable timber, purchasers should be indifferent to whether negatively valued timber is included in sales.

Mills that purchase timber from a few high-value forests — notably those in the Southern Piedmont, western Washington, western Oregon, parts of northern California, northern Idaho, and possibly northwest Montana — might not lose at all. In fact, a few of these forests — most likely those in northwest Oregon and western Washington — might even choose to depart from nondeclining flow and increase sales above the levels they will otherwise sell.

Ranchers who depend on public lands for a large share of their summer grazing might lose, particularly if they cannot afford to pay the full costs of grazing. Many ranchers have made improvements on public rangelands at their own expense, and the Forest Service should probably repay ranchers the depreciated value of their investments.

Timber Prices

The national forests provide a large — but not dominant — share of the timber in the United States. Since 1970, timber harvests from the national forests have averaged less than one-fourth of the nation's total wood supply. A 40 percent reduction in national forest sales will therefore represent less than a 10 percent reduction in the national wood supply.

Reduced national forest sales will actually have an even smaller effect on wood supply because other landowners will increase their harvests to partially compensate for these reductions. In 1980, the Congressional Budget Office (CBO) estimated that 40 to 60 percent of changes in national forest timber sales will be offset by compensating changes in private timber harvests. Changes in imports will offset another 20 to 30 percent of changes in national forest harvest levels. As a result, said the report, a 20 percent increase in national forest sales will produce less than a 1 percent increase in total domestic timber supply.[2]

The CBO also noted that end product prices "are influenced as much by processing and transportation costs as by costs of standing timber." Logging, road construction, and manufacturing costs averaged $200 per thousand board feet in 1977, far more than average national forest stumpage prices.[3] A decrease in national forest timber sales might increase stumpage prices, but it will not increase logging and manufacturing costs. Thus, the total price increase to consumers will be small.

As a result of all these factors, the Congressional Budget Office estimated that a change of 1.5 billion board feet in national forest timber sales will change national forest stumpage prices by less than 8 percent. A change of 3 billion board feet — just over 25 percent of current sales — will change stumpage prices by less than 25 percent. Ultimately, this will make only a 1 percent difference in the cost of a new home in 1990.[4] Although the CBO study was contemplating increases in national forest sales, similar effects will be expected with decreases in sales.

Similar figures were calculated by the Forest Service for the 1985 RPA Program. Economists estimated that a change of 1 billion board feet in timber supplies will change the cost of an average house by $250. If national forest sales fell by 5 billion board feet, the cost of a house might increase by $1,250 — which is still little more than 1 percent of the average cost of a new home.[5]

Predicting an impending timber shortage has become almost a full-time occupation for many analysts in the timber industry and Forest Service. A glossy advertisment in *Time* magazine paid for by the American Forest Institute warns that "timber demand [is] to increase faster than supply."[6] Yet there is no reason to expect that a shortage will ever take place. As C. W. Bingham, senior vice president of Weyerhaeuser Company, says, "There is *no* national timber supply shortage, either now or in the foreseeable future that cannot physically be met, *unless* public policy interferes with, or distorts, market economics to create a future shortage."[7]

Implementation of the proposals in chapter 14 will greatly reduce federal interference in the market for timber.

Forest Service Administration

As estimated above, this proposal will reduce Forest Service budgets by about 33 percent. Moreover, all of that money will be spent at the discretion of individual forests and not by regional and national offices. This will have a major effect on the structure of the Forest Service.

Out of a total of 30,650 permanent Forest Service employees, over 1,000 work in the Washington office. Many regional offices have over 300 staff members, and total regional and Washington office permanent staff members number close to 4,000. With funding for these offices provided only by contracts with the national forests, this proposal will probably reduce regional and Washington office staffing to much lower levels.

This is probably a good thing for everyone except the people whose jobs might be eliminated. When compared with sound examples of private enterprise, the Forest Service appears to have a disproportionate number of people in strictly administrative capacities. Nearly 15 percent of employees work at the national or regional offices. Such proportions are probably appropriate for a budget-maximizing agency that seeks continued funding from Congress. They are not appropriate for an organization or group of organizations that must be efficient.

The authors of a popular book on management, *In Search of Excellence*, note, "With rare exception, there is seldom need for more than 100 people in the corporate headquarters." As a typical example, they cite Emerson Electric, with 54,000 employees but fewer than 100 in the corporate headquarters. Another company, the Dana Corporation, reduced its corporate staff from 500 to 100 in 1973. Since that time, the company has grown more than 50 percent faster than American industry as a whole.[8]

The Research and State and Private Forestry branches of the Forest Service might also be threatened by this proposal. Yet the Research branch, which currently employs about 2,400 people, should be able to survive on contracts from chartered national forests as well as from other agencies. Very little Forest Service research is "basic" research; most relates to resources that national forests should be able to market. It will rarely be efficient for individual forests to conduct their own research, so they will find it expedient to jointly contract with experiment stations for research projects.

State and Private Forestry, whose staff numbers under 600, will survive mainly as a program for providing low-cost loans to private landowners. Given the power to issue tax-free bonds, the Forest Service should be able to obtain more favorable interest rates than small private landowners who wish to borrow money. Yet many private lands are more productive than the national forests. Chartered national forests should be able to encourage production on these lands by loaning their owners money at favorable rates of return.

National Forest Operations

A transition period of two to three years will be required to implement these proposed reforms. During this time, appropriations will be phased out and new fee schedules for recreation, grazing, and other resources phased in. Forest managers will work with the new boards of directors to develop strategies for each national forest, including methods of determining minimum bid prices for timber, recreation plans, and investments to be made with borrowed money and the seed money. After the transition period, each national forest will experiment with new methods of multiple-use management.

A typical forest might prepare marketing plans every five years. These plans will be tactical rather than strategic and will be based on market variables, new information about resource relationships, and changes in income and expenses. Because planning costs must be paid for out of income, the forest will not spend hundreds of thousands of dollars on computer runs, nor will the plans take five or more years to prepare.

The marketing plans will essentially identify opportunities to earn income over the next five-year period. Management of each resource will be evaluated for the compatibility with other resources, the risk that management will irreversibly damage forest resources, and the relationship between the five-year plan and sustained yield management.

A distinguishing factor of marketization will be the wide diversity of processes that individual national forests will probably adopt. Some will retain in-house staff for certain activities; others will contract those activities out. Some will form agreements with nearby forests and other resource agencies; others will find it best to work alone. The strategies that are most successful — that is, those that increase management budgets — are likely to be adopted by other forests where they are applicable. Because of this diversity, it is impossible to predict how any individual forest will respond to marketization. However, some possible tactics can be outlined for each forest resource.

Timber

Timber sales will be provided at levels that ensure that no sales lose money unless the losses are truly justified by another revenue-producing resource, such as recreation. Marginal forests, such as those in southern Idaho and North Carolina, might vary sale levels depending on the market for timber. Submarginal forests, such as those in Colorado and Alaska, will probably sell very little timber unless prices greatly increased. Forests with valuable timber, such as those in Mississippi and western Washington, will sell fairly consistent levels of timber unless prices drastically fell.

Nondeclining flow will not be an imperative, but managers will have incentives to practice sustained yield timber management. Since the forest keeps only a fraction of the income, supervisors will be disinclined to sell so much timber today that future income will ever be reduced to very low levels. Protection of wildlife, recreation, and scenic values — which contribute to other income sources — will also prevent large departures from nondeclining flow.

Individual timber sales will be designed to produce the highest net income. This might require clearcutting in many situations. However, the desire to reforest lands as cheaply as possible will lead many forests to experiment with other harvest systems, such as shelterwood cutting, which do not require universal hand planting. One way to reforest at no cost — in fact, at a profit — will be to offer stewardship contracts in which the people doing the reforestation will have the right to harvest the timber when it is mature. Such contracts will be freely transferable, so if the reforested land turned out to be more valuable for recreation in the future, the timber rights could be purchased and the timber not cut.

Under current Forest Service rules, neither timber rights nor grazing rights are freely transferable. Forests should be encouraged to allow timber purchasers, grazing permittees, recreation permit holders, owners of stewardship contracts, and other users to transfer their rights. This might create a speculative market for these resources, but it will transfer the risks of speculation away from timber purchasers and permittees toward professional speculators.[9]

Individual forests will probably contract timber inventory, soils survey, certain planning tasks, and other timber-related activities out to other organizations. In some cases, the regional offices could efficiently provide these skills, but they might have to compete with private organizations for business.

Recreation

Recreation will become the most important resource in most national forests and second only to timber in some forests in the Pacific Northwest, northern California, and the deep South. Forests will be able to experiment with various ways of collecting recreation fees, but some consistency might be required to reduce public confusion.

There are many possible methods of fee collection. An annual permit might give the holder and family members the right to several kinds of dispersed recreation throughout the year. Hunting and fishing permits will require higher fees. Daily or weekly permits will be available at a lower charge than an annual permit. Annual permits might be limited to dispersed recreation, while developed recreation will be paid for on a day-use basis.

Some forests might contract certain recreation acres to private developers or managers. This is already done for downhill skiing, but it could also be done for campgrounds, water sports, and perhaps even certain forms of dispersed recreation, such as fishing and hunting. Other forests might attempt to operate many of these activities themselves.

In general, forests will probably cooperate with other, nearby forests in establishing fees and issuing permits. One annual dispersed recreation permit might allow entry into several forests. At the same time, some forests or wilderness trusts might decide to provide premium forms of recreation on some of their lands. For example, one wilderness area might be managed for absolute primitive experiences: Holders of permits will be assured that they will meet no other permit holders during their trip. Naturally, such permits will be more expensive than most other wilderness permits.

Where nonrecreation activities take place, high recreation values will encourage forest managers to modify timber management or other developments in many areas. For example, marginal timber forests might gain recreation income if timber management were to cease completely over large parts of the forest. Forests with very valuable timber, such as those west of the Cascades, will continue to protect scenery, but otherwise timber values will be likely to swamp recreation values. Submarginal forests might spend a share of recreation income for a few small timber sales to open scenic vistas or for wildlife habitat improvement.

Wildlife

Most wildlife values will be expressed in the form of fees paid for wildlife recreation. Hunting and fishing fees in particular will give the Forest Service an incentive to protect game animals and fish. In addition, because nonconsumptive wildlife recreation represents a significant part of dispersed recreation, forests with a diversity of wildlife will be encouraged by recreation receipts to maintain that diversity.

Research results suggest that protection of habitat for some species of wildlife can benefit timber production. For example, timber managers will be encouraged to leave snags, which provide habitat for many insectivorous birds, as the least-cost method of preventing insect problems.

As previously described, threatened and endangered wildlife will continue to be protected under the Endangered Species Act. Funding for threatened and endangered species—perhaps from voluntary sources—will be channeled to the Fish and Wildlife Service. This agency might also contract with the Forest Service to protect some wildlife habitat above the levels strictly required by the Endangered Species Act.

Grazing

Grazing practices will change considerably under proposed reforms. Some base level of grazing can be produced at practically no cost in most forests. Yet many, if not most, grazing improvements are more costly than the benefits they provide. Grazing values might also be less than wildlife values, as evidenced by the hunting fees now being charged by private landowners in some areas.

Since forests must pay grazing costs out of a share of grazing income, they will be encouraged to sell grazing at fair market value. Some might sell grazing rights to the highest bidder. Others might maintain existing allotments but raise grazing fees to cover costs and gain a closer approximation of fair market value.

Watershed

Aside from compliance with clean water laws — which apply to private forestland owners as well as the Forest Service — several things might encourage the Forest Service to protect water quality. In some forests, fish-related fees will exceed the value of timber or grazing activities which might reduce water quality and therefore fish income.

Perhaps more important, many forests include municipal watersheds. The forests could contract with municipalities to provide high-quality water. An important legal question is who owns the water? If the national forest owns the water, then the municipality may have to pay the Forest Service to protect water quality. If the city is the owner, then the Forest Service may have to pay the municipality if water quality is reduced. This question may be resolvable according to previous congressional acts setting aside certain watersheds for the use of various cities.

Either way the question is answered, the outcome is likely to be the same. In some watersheds, the value of timber is so small that the benefits from timber sales will be outweighed by the costs of filtration plants, which will be installed to maintain water quality. In others, timber values might outweigh the costs of filtration. Whoever pays, the filtration plants will be installed in the latter case, while timber management will be reduced in the former.

Table 15.1 Montana Roadless Areas Scheduled for Timber Sales

Forest	*Acres Scheduled for Timber Sale*	*Acres Reforms Are Likely to Protect*
Beaverhead	150,000	150,000
Bitterroot	173,200	173,200
Custer	72,100	72,100
Deerlodge	161,500	161,500
Flathead	29,675	29,675
Gallatin	118,700	108,900
Helena	70,300	70,300
Kootenai	93,500	23,375
Lewis & Clark	67,200	67,200
Lolo	152,900	152,900
Total	1,168,300	1,009,150

More than 1.1 million acres of Montana roadless lands are scheduled for development in the next 10 to 20 years by Forest Service plans. CHEC estimates that marketization will protect virtually all of the roadless areas in most of these forests. The major exceptions will be in the Kootenai and Flathead Forests, where timber is more valuable; marketization will probably still protect substantial portions of the roadless lands in these forests.

Source: Forest plans for the indicated forests.

Wilderness

Most roadless areas which have not yet been designated wilderness by Congress are being scheduled for development by the Forest Service, which is planning to build about 700 to 900 miles of roads per year into roadless areas during the next 10 years.[10] The annual cost of constructing these roads will be $34 to $44 million. The proposed reforms will greatly change these plans, particularly in the regions where timber sales are often below-cost and recreation values are greater than commodity values.

For example, forest plans for the 10 Montana national forests indicate that nearly 1.2 million acres of roadless lands are to be developed for timber sales in these forests (table 15.1). Most of these forests lose money on timber sales, particularly when road construction is required. It is unlikely that the costs of developing these roadless areas would ever be recovered by taxpayers, yet the value of the roadless areas for recreation is great. CHEC estimates that slightly over 1 million acres of these lands will be protected by the proposed reforms.

Winners and Losers

These proposals represent drastic changes in public forest management, not to mention some touchy changes in western water law. Yet there are good reasons to expect that these changes are possible.

Environmentalists should support the changes because the changes will result in a clear improvement in national forest management and have a good chance of improving private forest management as well. Some conflicts between timber and endangered wildlife will remain, but most other environmental battles in the national forests will be resolved.

Counties should support the changes because national forest payments to counties will increase. Cross-subsidized timber sales reduce county payments by $12.5 million per year or more. The failure to collect reasonable recreation fees reduce payments even more. As estimated above, this proposal will increase county income by hundreds of millions of dollars per year.

Many private landowners should support the changes because they will provide new sources of income and eliminate controversy over who owns land resources. Currently, some state legislatures are considering the passage of laws giving people free access to hunting and fishing on private lands. Such laws will encourage landowners to abuse fish and wildlife habitat so the landowners will not be bothered with what they regard as trespassers.

Private timber companies and ranchers may or may not oppose these proposals, depending on their particular situation. The public land agencies might be so wedded to the status quo that they will oppose reductions in their bureaucracy no matter how beneficial such reductions will be to the public. But with the broad base of support described here, reforms should be possible in spite of such opposition.

Notes

1. "Total Benefits – 1985 RPA High Bound Program," unpublished document on file at Forest Service office in Washington, DC, 1987.
2. Lawrence H. Oppenheimer, *Forest Service Timber Sales: Their Effect on Wood Product Prices* (Washington, DC: Congressional Budget Office, 1980), p. xiv.
3. Ibid., p. 20.
4. Ibid., p. xv.
5. USDA Forest Service, *A Recommended Renewable Resource Program: 1985–2030* (Washington, DC: Forest Service, 1986), p. 3.
6. American Forest Institute, *Greater Expectation: The American Forest's Potential* (Washington, DC: American Forest Institute, 1981), p. 3.
7. Charles W. Bingham, "Defining the Timber Supply Concerns of the Timber-Growing Industry," presentation before the Forest Products Research Society, 2 October 1979, p. 1.
8. Thomas J. Peters and Robert H. Waterman, Jr., *In Search of Excellence* (New York, NY: Harper & Row, 1982), pp. 272–273.
9. Barney Dowdle, "Reforming Forest Service Timber Sales," *Forest Watch* 6(11):15–17.
10. Letter from Chief R. Max Peterson to Lynn Greenwalt, National Wildlife Federation, 26 March 1985, 4 pp.

Conclusion
A New Environmental Agenda

Traditional environmental assessments of national forest controversies are based on a fundamental misdiagnosis of the problem. Correcting this misdiagnosis leads to a new view of the Forest Service: Not bad people intent on environmental destruction and economic waste, but ordinary people motivated by incentives, like everyone else.

The appropriations process has taught the Forest Service that Congress is most responsive to production of commodities such as timber and grazing. Laws like the Knutson-Vandenberg Act and the Public Rangelands Improvement Act also encourage commodity production. Below-cost timber sales, overgrazing, road construction in potential wilderness areas, and neglect of important recreation values, as well as fabrication of data in forest plans or reforestation reports, all result from the agency's desire to maximize its budget.

If this is true, then replacing the people who run the Forest Service, perhaps through presidential elections, will have negligible effects on the agency — as John Crowell learned to his frustration. For example, if all Forest Service foresters were replaced by wildlife biologists tomorrow, the biologists would soon propose below-cost sales to augment K-V funds for wildlife. Changing the legal prescriptions under which the agency must operate will merely lead to more double-talk and clever accounting systems. Reform of the Forest Service must come by changing the incentives that motivate national forest managers.

This necessarily means divorcing the Forest Service from congressional appropriations. Appropriations are politically determined through a process that favors those special interest groups that can gain the most income. In the long run, national forest appropriations will be oriented toward development and environmental destruction.

Changing incentives also requires charging for recreational use. Managers whose budgets increase when recreational use increases but decrease when below-cost sales occur will act very differently from the way Forest Service officials act today. Any reforms that are not accompanied by recreation fees will fail to protect national forest amenity values.

It is true that sacrifices will be needed to achieve these reforms. Recreationists must accept increased recreation fees, but in exchange they will gain a significant increase in recreation opportunities. Environmentalists must accept the fact that some national forests will be managed predominantly for timber, but in exchange they will halt development of millions of acres of below-cost timber lands. Timber purchasers must accept that many national forests will sell virtually no timber, but in exchange they will gain a stabilization of timber supplies throughout the National Forest System. The Forest Service must accept a reduction in its bureaucracy, but in exchange it will see an end to morale problems caused by polarization, red tape, and delays.

Marketization may appear revolutionary, but unlike many revolutionary ideas that are politically infeasible, marketization can be successfully promoted by a coalition of interest groups simply because so many interests will benefit. These beneficiaries include:

- Recreationists, who will gain expanded opportunities for hunting, fishing, camping, and other outdoor recreation activities;
- Counties, whose incomes from national forest management are likely to double or triple;
- Private landowners, whose timber, forage, and recreation resources will all gain in value when below-cost sales cease in the national forests;
- Wilderness advocates and wildlife lovers who are concerned about the effects of below-cost timber sales and grazing on wildlands, water quality, and other values;
- Forest managers who are tired of the pointless red tape of environmental impact statements, environmental analyses, and other paperwork; and
- Fiscal conservatives and others who are concerned about the federal deficit.

Perhaps the largest and most powerful of these interest groups is the environmental community, which has been fairly evenly matched with the timber industry in previous congressional debates. In 1976, for example, the timber industry and the Forest Service were able to convince Congress to reject most environmental reforms only because wildlife interest groups sided with the industry and agency. With the support of wildlife groups, environmentalists may have veto power over industry-supported legislation — but the industry probably would retain veto power over legislation supported only by environmentalists.

Thus, the support of counties, private landowners, and other interest groups is critical to the passage of Forest Service reforms. These groups, and particularly the counties, traditionally support the timber industry because they are not aware of what inefficient national forest management is costing them. The proposed reforms offer them many benefits and should be politically feasible if they side with the environmental community.

Environmentalists have much to gain from reforms, and reforms are not likely to take place at all without their support. Yet the dramatic changes proposed by this book will at first seem risky to most environmentalists. The dominant environmental paradigm — that recreation, wildlife, water, and other amenities are strictly nonmarket resources that will be destroyed without government regulation and prescriptive legislation — is not easy to subvert.

Yet marketization is far superior to the current system of interest group politics. It will protect millions of acres of wildlands that the Forest Service plans to road and log. It will stop below-cost pesticide spraying and bring overgrazing under control. It will provide a wider variety of recreation opportunities than is available today. It will put an end to fighting losing battles against environmentally destructive and economically absurd projects. In short, the proposals in this book should provide a blueprint for environmental action regarding the Forest Service and many other natural resource agencies.

Index

A

B

E

F

G

H

I

J

K

T

U

V

W

Y

About the Author

Randal O'Toole is a forest economist with Cascade Holistic Economic Consultants (CHEC), a forestry consulting firm in Eugene, Oregon. Working primarily as a consultant to environmental groups, O'Toole has prepared detailed analyses of over 40 national forest plans. These reviews, combined with local public involvement in the planning process, led the Forest Service to reduce the amount of land it manages for timber by over 1 million acres in Alaska's Chugach National Forest, 600,000 acres in New Mexico national forests, and many more acres in other forests from Montana to North Carolina.

Since 1980, O'Toole has been the publisher of *Forest Watch* (formerly *Forest Planning*) magazine, a monthly periodical designed to help people understand and influence public forest management. To further help people comprehend forest issues, O'Toole has also authored or co-authored handbooks such as the *Citizens' Guide to Timber Management* and the *Citizens' Guide to Forestry and Economics*. O'Toole has also done original research on a number of topics, focusing on the economics of problems such as reforestation, wilderness, old-growth, and timber sales. Copies of these research papers and other publications are available from CHEC, P.O. Box 3479, Eugene, Oregon 97403.

Also Available from Island Press

Americans Outdoors: The Report of the President's Commission—The Legacy, The Challenge, with case studies
Foreword by William K. Reilly
1987, 426 pp., appendixes, case studies, charts
Paper: $24.95 ISBN 0-933280-36-X

The Challenge of Global Warming
Edited by Dean Edwin Abrahamson
Foreword by Senator Timothy E. Wirth
In cooperation with the Natural Resources Defense Council
1989, 350 pp., tables, graphs, index, bibliography
Cloth: $34.95 ISBN: 0-933280-87-4
Paper: $19.95 ISBN: 0-933280-86-6

Crossroads: Environmental Priorities for the Future
Edited by Peter Borrelli
1988, 352 pp., index
Cloth: $29.95 ISBN: 0-933280-68-8
Paper: $17.95 ISBN: 0-933280-67-X

Down by the River: The Impact of Federal Water Projects and Policies on Biodiversity
By Constance E. Hunt with Verne Huser
In cooperation with the National Wildlife Federation
1988, 256 pp., illustrations, glossary, index, bibliography
Cloth: $34.95 ISBN: 0-933280-48-3
Paper: $22.95 ISBN: 0-933280-47-5

Forest and the Trees: A Guide to Excellent Forestry
By Gordon Robinson
Introduction by Michael McCloskey
1988, 272 pp., indexes, appendixes, glossary, tables, figures
Cloth: $34.95 ISBN: 0-933280-41-6
Paper: $19.95 ISBN: 0-933280-40-8

Holistic Resource Management
By Allan Savory
Center for Holistic Resource Management
1988, 512 pp., color plates, diagrams, references, notes, index
Cloth: $39.95 ISBN: 0-933280-62-9
Paper: $24.95 ISBN: 0-933280-61-0

Land and Resource Planning in the National Forests
By Charles F. Wilkinson and H. Michael Anderson
Introduction by Arnold Bolle
1987, 400 pp., index
Paper: $19.95 ISBN: 0-933280-38-6

Last Stand of the Red Spruce
By Robert A. Mello
Introduction by Senator Patrick J. Leahy
In cooperation with the Natural Resources Defense Council
1987, 208 pp.
Photographs, charts, index
Paper: $14.95 ISBN: 0-933280-37-8

Natural Resources for the 21st Century
Edited by R. Neil Sampson and Dwight Hair
In cooperation with the American Forestry Association
1989, 350 pp., index, illustrations
Cloth: $39.95 ISBN: 1-55963-003-5
Paper: $19.95 ISBN: 1-55963-002-7

Our Common Lands: Defending the National Parks
Edited by David J. Simon
Foreword by Joseph L. Sax
In cooperation with the National Parks and Conservation Association
1988, 575 pp., index, bibliography, appendixes
Cloth: $45.00 ISBN: 0-933280-58-0
Paper: $24.95 ISBN: 0-933280-57-2

Saving The Tropical Forests
By Judith Gradwohl and Russell Greenberg
Preface by Michael H. Robinson
Smithsonian Institution
1988, 207 pp., index, tables, illustrations, notes, bibliography
Cloth: $24.95 ISBN: 0-933280-81-5

Shading Our Cities: Resource Guide for Urban and Community Forests
Edited by Gary Moll and Sara Ebenreck
In cooperation with the American Forestry Association
1989, 350 pp., index, illustrations, photographs, appendixes
Cloth: $34.95 ISBN: 0-933280-96-3
Paper: $19.95 ISBN: 0-933280-95-5

These titles are available directly from Island Press, Box 7, Covelo, CA, 95428. Please enclose $2.75 shipping and handling for the first book and $1.25 for each additional book. California and Washington, DC, residents add 6% sales tax. A catalog of current and forthcoming titles is available free of charge.